Lecture Notes in Mathematics

Edited by A. Dold and B. Eckmann

1218

Schrödinger Operators, Aarhus 1985

Lectures given in Aarhus, October 2–4, 1985

Edited by E. Balslev

Springer-Verlag
Berlin Heidelberg New York London Paris Tokyo

Editor

Erik Balslev
Matematisk Institut, Aarhus Universitet,
NY Munkegade, Bygning 530
8000 Aarhus C, Denmark

Mathematics Subject Classification (1980): 35J, 81F

ISBN 3-540-16826-5 Springer-Verlag Berlin Heidelberg New York
ISBN 0-387-16826-5 Springer-Verlag New York Berlin Heidelberg

Printing and binding: Druckhaus Beltz, Hemsbach/Bergstr.
2146/3140-543210

CONTENT

Introduction.

The present volume is a collection of papers based on lectures delivered at a Symposium on Schrödinger operators held at the Institute of Mathematics, Aarhus University, october 2nd - 4th 1985. The speakers presented recent results on a fairly wide range of problems.

A paper with more than one author was presented by the first listed author.

We want to thank the Danish Natural Science Foundation for financial support.

Aarhus, June 1986.

Erik Balslev.

The Schrödinger operator for a particle in a solid with deterministic and stochastic point interactions

by

S. Albeverio[**,*], F. Gesztesy[***], R. Høegh-Krohn[#], H. Holden[##], W. Kirsch[*]

[*] Fakultät für Mathematik, Ruhr-Universität, D-4630 Bochum 1

[**] Research Centre Bielefeld-Bochum Stochastics (BiBoS)

[***] Laboratoire de Physique Théorique, Université des Paris-Sud, Orsay; on leave of absence from Institut für Theoretische Physik, Universität Graz

[#] Matematisk Institutt, Universitetet i Oslo

[##] Courant Institute of Mathematical Sciences, New York; on leave of absence from #

ABSTRACT

We give a survey of recent results concerning Schrödinger operators describing the motion of a quantum mechanical particle in $\mathbb{R}^3$ or $\mathbb{R}^1$ under the influence of a potential concentrated at N centers, $N \leq \infty$.
We dedicate particular attention to the case $N = \infty$, with centers forming a periodic lattice (model of a crystal) or with centers randomly distributed with random strengths (models of disordered solids or random alloys).

0. Introduction

The study of the motion of quantum mechanical particles in an ordered or disordered solid presents formidable difficulties and various simplifications/idealizations have to be made. In the "one electron approximation" just one particle (electron) is considered and the centers of forces (nuclei) are fixed, belonging to a lattice (nuclear vibrations and presence of other electrons are thus neglected).

In the case where the total interaction potential between the electron and the nuclei is periodic ("Bloch/Floquet model"), structural results (band structure, estimates on number of gaps) on the spectrum are known, see e.g. [1], [2], [3], [4], [5]. In the case of randomly disturbed lattices also some general results are known, see e.g. [6], [7], [8]. However in both cases there is great interest in having "solvable models" in which all quantities can be calculated. Such models have also the important function of dealing as reference for testing approximate mathematical methods.

An important class of solvable models is provided by the so called point interaction models (also known under the name of δ-interaction models or zero range models; they are also closely related to "pseudo potentials" models). These models have been introduced in solid state physics and nuclear physics since the thirties, see [9] (which surveys up to 65 work on the 1-dimensional case), [10], [11], [12]. One should also mention later applications to problems of electromagnetic theory, see e.g.[13,14]. In particular the Kronig-Penney model of a one-dimensional crystal with periodic δ-interactions is well known and have played an important role as a reference model in solid state physics, see e.g. [15]. The 3-dimensional analogue of this model has only been brought under mathematical control in recent years (see e.g.[11], [16,17,34,40] and references therein) and should come to play an important role also for more realistic approaches of solid state physics. The first part of each section of this paper is dedicated to a description of this model and of the corresponding results concerning the spectrum.

For the description of disordered solids, like those arising from impurities or from alloys, models with stochastic interactions have been investigated, in particular in recent years, under the stimulus of Anderson-Mott discussion of the phenomenon of localization; there is by now a quite large literature, see e.g.[8], [18],[11] and references therein. Also in this case, the study of solvable models provides new insights, as was already clear since the work around the Hutton-Saxner conjecture in one-dimension (see e.g. [9] and references therein). Recently the one-dimensional case of models of alloys built with δ-interaction of random strength and position has been discussed in [6], [7], [19]. The case of three dimensional models with δ-interactions of random strengths and positions has also been discussed recently, see [20]-[23] . We take up this subject in Sect. 2. Finally we also mention some other models involving interactions localized at certain subsets of $\mathbb{R}^3$, like those involving δ-shell models, models of electrons interacting with polymers and models for self-interacting polymers, see e.g. [24] resp.25, 29] and references therein.

1. Schrödinger operators with non random point interactions

Let us start by explaining the meaning of the words "point interactions" in the title of this lecture. By this one understands in general, in the theory of Schrödinger operators, hence in quantum theory, the presence in the Hamiltonian of potentials, "interactions", of the form of operators multiplication by a "function" V supported on some subset of $\mathbb{R}^d$ consisting of isolated points (or, more generally, of measure zero). "Function" is to be understood in some generalized sense (even more general than distribution), and the first problem one meets is of course to define properly say $-\Delta + V$ as an operator in $L^2(\mathbb{R}^d, dx)$, with Δ the Laplacian in $\mathbb{R}^d$.

To illustrate what we have mind let us consider a couple of typical examples.

a) $d \leq 3$, $-\Delta + \lambda\delta(x)$. This model has a long history in nuclear physics, as an idealized model of a short range force (Bethe-Peierls, Thomas 1935; see e.g. [11] [12] [26]).

b) $d \leq 3$, $-\Delta + \sum_{y \in Y} \lambda_y \delta(x-y)$ for Y a discrete subset of $\mathbb{R}^d$. This model has a long history in solid state physics, at least for d = 1 and Y periodic, e.g. $\mathbb{Z}$: "Kronig-Penney" model for the motion of an electron in an idealized crystal (see e.g. [15], [11]).

c) $-\Delta + \lambda V(x)$ with $|\mathrm{supp}\, V| = 0$ ($|\ |$ meaning Lebesgue measure) (e.g. supp V the surface of a sphere for d = 3: "delta shell model", see e.g. [24], [27]).

d) $-\Delta + \lambda\int_0^t \delta(x-b(s))ds$, with $\{b(s),\ 0 \leq s \leq t\}$ a continuous curve in $\mathbb{R}^d$, like e.g. a path of a Brownian particle (this latter case constitutes a model of a quantum mechanical particle interacting with a polymer, see e.g. [25],[28],[29]).

The questions one can ask are: can one associated with above heuristic Hamiltonians well defined self-adjoint operators H in $L^2(\mathbb{R}^d)$?
What about their spectrum, eigenfunctions etc.?

1.1 The case of a one center interaction in $\mathbb{R}^3$

Let us first recall some methods for defining H in the case a). The first observation is that the case $d \geq 4$ is trivial in as much as formally $H = -\Delta$ on $C^\infty_{o,o}(\mathbb{R}^d)$ (the C^∞ functions of compact support vanishing in a neighborhood of zero) and $-\Delta \restriction C^\infty_{o,o}(\mathbb{R}^d)$ is essentially self-adjoint for $d \geq 4$, cfr. e.g. [30], Ch. X), hence there is no self-adjoint extension of $H \restriction C^\infty_{o,o}(\mathbb{R}^d)$ different from the trivial one, $-\Delta$.

We shall now examine in some details the most interesting case for physics, namely d = 3 (for the cases d = 1,2 see [11]).

a) 1. The first method for defining $H = -\Delta + \lambda\delta(x)$ in $L^2(\mathbb{R}^3)$ is to use nonstandard analysis. We shall not describe here this method (see [31], [32], [25], [11]), we just give a hint. Let ε be a positive infinitesimal. By transfer $H_\varepsilon \equiv -\Delta + \lambda_\varepsilon \delta_\varepsilon$ is well defined, self-adjoint in ${}^*L^2(\mathbb{R}^3)$ with $\delta_\varepsilon(x) = (\frac{4}{3}\pi\varepsilon^3)^{-1} \chi_1(|x|/\varepsilon)$, a non standard realization of the δ-function, χ_1 being the characteristic function of a sphere of radius 1 and centre at the origin, and

$$\lambda_\varepsilon = -\frac{2}{3}\pi^3\varepsilon + \frac{(4\pi)^2}{3}\alpha\varepsilon^2, \ \alpha \in \mathbb{R}.$$

Then the standard part (defined by the resolvent) of H_ε exists and is equal to the self-adjoint operator $-\Delta_\alpha$ describing a "point interaction of strength α at the origin" (we shall give below a standard description of $-\Delta_\alpha$). This non standard procedure shows also that the only possible realizations of "$-\Delta + \lambda\delta(x)$" form a one-parameter family, with parameter $\alpha \in \mathbb{R}$. In fact also α a positive infinite number is allowed, in which case $-\Delta_\alpha$ is simply the free Hamiltonian $-\Delta$. It also indicates that an infinitesimal negative coupling constant is needed to obtain, after taking standard parts, an Hamiltonian different from $-\Delta$.

It also turns out that the sign of the "renormalized coupling constant" α determines whether or not $-\Delta_\alpha$ has a bound state (eigenvalue) (at $-(4\pi\alpha)^2$), namely if $\alpha < 0$ there is an eigenvalue, if $\alpha \geq 0$ there is no eigenvalue.

a) 2. Let us now give shortly a first standard method of constructing $-\Delta_\alpha$, for details and proofs see [33], [11]. Let V be Rollnik (i.e. $\iint |V(x)|\,|V(y)|\,|x-y|^{-2}\,dxdy < \infty$) and in $L^1(\mathbb{R}^3)$, and let λ be $C^1(\mathbb{R})$ with $\lambda(0) = 1$. We shall study $-\Delta + \lambda(\varepsilon)\,\varepsilon^{-2}\,V(x/\varepsilon)$ as $\varepsilon \downarrow 0$. Set $v \equiv |V|^{1/2}$, $u \equiv v \operatorname{sign} V$, $G_k(x) \equiv \frac{e^{ik|x|}}{4\pi|x|}$ for $\operatorname{Im}k > 0$, $x \neq 0$, $x \in \mathbb{R}^3$ (so that $G_k(x) = (-\Delta - k^2)^{-1}(x)$ is the kernel of the resolvent of $-\Delta$).

If $\varphi \in L^2(\mathbb{R}^3)$ solves $uG_0 v\varphi = -\varphi$ then define $\psi \equiv G_0 v\varphi$. Then from the assumption on V we have $\psi \in L^2_{loc}(\mathbb{R}^3)$, $\nabla\psi \in L^2(\mathbb{R}^3)$, $H\psi = 0$ in the sense of distributions, where $H \equiv -\Delta + V$ (in the sense of quadratic forms). If $\psi \notin L^2(\mathbb{R}^3)$ then one says that H has a zero energy resonance (or there is such a resonance for V). If $\psi \in L^2(\mathbb{R}^3)$ then $H\psi = 0$ in sense that $\psi \in D(H)$ (the definition domain of H) and $H\psi = 0$. In this case ψ is a zero energy eigenstate for H. One has, under the additional assumption $|x|\,V(x) \in L^1(\mathbb{R}^3)$: $\psi \in L^2(\mathbb{R}^3) \Leftrightarrow (v,\varphi) = -\int V\psi dx = 0$.

The following case distinction is important (we assume here $(1 + |\cdot|)\,V \in L^1(\mathbb{R}^3)$):

Case I: -1 is not an eigenvalue of $uG_0 v \equiv K$ (this is the case when $V \geq 0$, e.g.).

Case II: -1 is a simple eigenvalue of K and $\psi \notin L^2(\mathbb{R}^3)$ i.e. ψ is not an eigenfunction (i.e. there is a "simple zero energy resonance").

Case III: -1 is an eigenvalue of K with eigenfunctions φ_j and the corresponding ψ_j are in $L^2(\mathbb{R}^3)$ for all j (in this case there is no zero energy resonance).

Case IV: -1 is an eigenvalue of K, with eigenfunctions φ_j and at least one $\psi_j \notin L^2(\mathbb{R}^3)$.

Under the above assumption on the potential and on λ, and in addition assuming $\lambda'(0) \neq 0$ in cases III, IV, we have that $n - \lim_{\varepsilon\downarrow 0} (H_\varepsilon - k^2)^{-1} = (-\Delta_\alpha - k^2)^{-1}$, where n-lim means norm-limit and $\alpha = +\infty$ in cases I, III (so that in these cases $-\Delta_\alpha = -\Delta$). In case II we have $\alpha = -\lambda'(0)\ |(v,\varphi)|^{-2}$; in case IV we have $\alpha = -\lambda'(0)\{\sum_{\ell=1}^{N} |(v,\varphi_\ell)|^2\}^{-1}$ (in these formulae we have chosen suitable normalizations of the φ, φ_ℓ e.g., in case II, (sign $V\varphi,\varphi$) $= -1$, where $(,)$ is the scalar product in $L^2(\mathbb{R}^3)$).

Remark: The intuitive reason for the dependence of α on the zero energy behaviour of $-\Delta + V$ can be found in the fact that

$$H_\varepsilon = \varepsilon^{-2} U_\varepsilon(-\Delta + \lambda(\varepsilon)V(\cdot))U_\varepsilon^{-1},$$

where U_ε is the unitary scaling $(U_\varepsilon g)(x) = \varepsilon^{-3/2} g(x/\varepsilon)$, $\varepsilon > 0$ in $L^2(\mathbb{R}^3)$, hence $\sigma(H_\varepsilon) = \varepsilon^{-2}\sigma(-\Delta + \lambda(\varepsilon)V(\cdot))$, where $\sigma(\cdot)$ means spectrum. This indicates that the behaviour of $\sigma(H_\varepsilon)$ as $\varepsilon\downarrow 0$ is determined by the behaviour at 0 of $\sigma(-\Delta + \lambda(0)V(\cdot))$.

Remark: The above procedure is a construction of the point interaction Hamiltonian $-\Delta_\alpha$ as limit of Schrödinger operators with "scaled local potentials". Other regularizations of the interaction have been discussed in the literature, see e.g. [51],[11] (and references therein).

a) 3. Perhaps the most directly usable characterization of $-\Delta_\alpha$ is by its resolvent kernel:

$$(-\Delta_\alpha - k^2)^{-1} = G_k - (\overline{G_k(\cdot)}, \cdot) A G_k, \tag{1.1}$$

with $A \equiv B^{-1}$, $B \equiv (\frac{ik}{4\pi} - \alpha)$, $\qquad$ $\operatorname{Im} k > 0$, $k^2 \notin \sigma(-\Delta_\alpha)$,

$$\text{where } \sigma(-\Delta_\alpha) = \begin{cases} [0,\infty) & \text{for } \infty \geq \alpha \geq 0 \\ \{-(4\pi\alpha)^2\} \cup [0,\infty) & \text{for } \alpha < 0. \end{cases}$$

It follows easily $\sigma_{ess}(-\Delta_\alpha) = \sigma_{ac}(-\Delta_\alpha) = [0,\infty)$, $\sigma_{sc}(-\Delta_\alpha) = \emptyset$.

For $\alpha < 0$ the normalized eigenfunction φ to the simple eigenvalue $-(4\pi\alpha)^2$ is $(-\alpha)^{1/2}\ |x|^{-1} \exp(4\pi\alpha\ |x|)$. For $\alpha > 0$, there is a resonance at $k = 4\pi i \alpha$.

a) 4. Take $\varphi_\alpha = |\alpha|^{1/2}|x|^{-1} \exp(4\pi\alpha|x|)$ for $x \neq 0$, $\alpha \in \mathbb{R}$, $\varphi \equiv 1$ for $\alpha = +\infty$. The quadratic form $f \in C_0^\infty(\mathbb{R}^3) \to \int(\nabla f)^2\ \varphi_\alpha^2 dx$ in $L^2(\mathbb{R}^3, \varphi_\alpha^2 dx)$ is well defined and closable. The self-adjoint positive operator $\hat{H}_\alpha$ in $L^2(\mathbb{R}^3, \varphi_\alpha^2 dx)$ associated with its closure is unitary equivalent $-\Delta_\alpha + (4\pi\alpha)^2$ for $\alpha \in \mathbb{R}$, in fact $\hat{H}_\alpha = \varphi_\alpha^{-1}[-\Delta_\alpha + (4\pi\alpha)^2]\varphi_\alpha$. For $\alpha = +\infty$ it is $-\Delta$.

The interest of the realization $\hat{H}_\alpha$ is that $\hat{H}_\alpha = -\Delta - \beta_\alpha \cdot \nabla$ on $C_0^\infty(\mathbb{R}^3)$-functions in $L^2(\mathbb{R}^3, \varphi_\alpha^2 dx)$, hence with $\beta_\alpha \equiv \nabla \ln\varphi_\alpha$. This is a "diffusion operator", whose closure generates a Markov semigroup in $L^2(\mathbb{R}^3, \varphi_\alpha^2 dx)$, with invariant measure $\varphi_\alpha^2 dx$ (a probability measure for $\alpha < 0$). $\hat{H}_\alpha$ is the operator associated with the closed Dirichlet form $\int(\nabla f)^2\ \varphi_\alpha^2 dx$ in $L^2(\mathbb{R}^3, \varphi_\alpha^2 dx)$. Thus $\hat{H}_\alpha$ has an immediate probabilistic meaning as generator of a symmetric diffusion process (whereas $-\Delta_\alpha + (4\pi\alpha)^2$ has only the interpretation of a Brownian motion disturbed by "creation of mass" at the origin). For these concepts and results see [35], [36], [37].

1.2 The case of point interactions at N centers

Of course there was nothing special in choosing the source in the origin, we could have treated on the same footing the operator $-\Delta + \lambda\delta_y(x)$ for any $y \in \mathbb{R}^3$.

In fact all considerations extend also to $-\Delta + \sum_{y\in Y} \lambda_y \delta_y(x)$, with Y any finite subset of $\mathbb{R}^3$. We shall limit ourselves here to give the formula for the resolvent.

Similarly as for the case $Y = \{0\}$ treated above, each coupling constant λ_y has to be renormalized to yield a corresponding $\alpha_y \in \mathbb{R} \cup \{+\infty\}$. Denoting by $-\Delta_{\alpha,Y}$ the realization of $-\Delta + \sum_{y\in Y} \lambda_y \delta_y(x)$ as a well defined self-adjoint operator on $L^2(\mathbb{R}^3, dx)$ we get correspondingly to a 3) above:

$$(-\Delta_{\alpha,Y} - k^2)^{-1} = G_k - \sum_{y,y'\in Y} (\overline{G_k(\cdot - y)}, \cdot)\, A_{yy'}\, G_k(y' - \cdot), \tag{1.2}$$

with $k^2 \notin \sigma(-\Delta_{\alpha,Y})$, $\mathrm{Im}\, k > 0$,

$A_{yy'}$, the kernel of the operator A in $\ell^2(Y)$ defined by $A = B^{-1}$, where B is the operator in $\ell^2(Y)$ with kernel $B_{yy'} \equiv (\frac{ik}{4\pi} - \alpha_y)\, \delta(y - y') + \tilde{G}_k(y-y')$, $\tilde{G}_k(z) \equiv G_k(z)$ for $z \neq 0$, $\tilde{G}_k(z) = 0$ for $z = 0$.

For $Y = \{0\}$, $\alpha_o = \alpha$, $A_{yy'} = A$ this reduces to the formula in a) 3. i.e. in this case $-\Delta_{\alpha,Y} = -\Delta_\alpha$.

Remark: As discussed in [38] for $|Y| = n$ there are actually n^2 self-adjoint extensions of $-\Delta \restriction C^\infty_{o,o}(\mathbb{R}^3)$, where the first zero means compact support and the second vanishing at a neighborhood of Y. The above n parametric realization $-\Delta_{\alpha,Y}$ can be shown to be the one given by separated boundary conditions at each point, see [11,38,39].

Remark: The same results for σ_{ess}, σ_{ac}, σ_{sc} as for $-\Delta_\alpha$ hold. The point spectrum of $-\Delta_{\alpha,Y}$ is entirely contained in $(-\infty, 0)$ and consists of at most N eigenvalues counting multiplicity and the eigenvalues are given by k^2 with $\mathrm{Im}\, k > 0$ s.t. $\det B(k) = 0$, the multiplicity of the zero eigenvalue of $B(k)$ being equal to the multiplicity of the eigenvalue k^2. See e.g. [11], where also many other results on point interactions with N centers can be found.

We shall now procede to the most interesting case for us, the case of infinitely many centers.

1.3 Point interactions at a discrete set of centers

We shall consider heuristic Hamiltonians of the form $-\Delta + \sum_{y\in Y} \lambda_y \delta_y(\cdot)$, with Y a discrete infinite set i.e. $Y \equiv \{y_j \in \mathbb{R}^3 \mid j \in \mathbb{N},\ \inf_{j\neq j'} |y_j - y_{j'}| > 0$.

In analogy with our discussion in 1.1, 1.2 we shall have to define properly an

Hamiltonian $-\Delta_{\alpha,Y}$ "realizing" the above heuristic one, by "renormalizing" the coupling constants λ_y to α_y with α a real-valued function on Y (as we know from 1.1, 1.2 the value $+\infty$ of α_y would simply correspond to deleating this y from Y).

We shall give a description of $-\Delta_{\alpha,Y}$ by its resolvent, using the one given in 1.2. Let thus $\tilde{Y}$ run over the finite subsets of Y and let $\tilde{\alpha} \equiv \alpha \restriction \tilde{Y}$. Then, by 1.2, $(-\Delta_{\tilde{\alpha},\tilde{Y}} - k^2)^{-1}$ is the well defined resolvent of a lower bounded self-adjoint operator $-\Delta_{\tilde{\alpha},\tilde{Y}}$. The following theorem can be proven, see e.g. [11]:

<u>Theorem 1</u>: For Im $k^2 \neq 0$ the strong limit as $\tilde{Y} \uparrow Y$ of $(-\Delta_{\tilde{\alpha},\tilde{Y}} - k^2)^{-1}$ exists and is the resolvent $(-\Delta_{\alpha,Y} - k^2)^{-1}$ of a self-adjoint operator. This resolvent is given by

$$(-\Delta_{\alpha,Y} - k^2)^{-1} = G_k + \sum_{j,j'=1}^{\infty} [\Gamma_{\alpha,Y}(k)]^{-1}_{j,j'} |G_k(\cdot - y_{j'})\rangle\langle \bar{G}_k(\cdot - y_j)| \qquad (1.3)$$

$\Gamma_{\alpha,Y}(k)$ is the closed operator in $\ell^2(Y)$ given by $(\Gamma_{\alpha,Y}(k)) = -(B_{y_j,y_{j'}})^{-1}$, with $B_{y_j y_{j'}}$ defined as in 1.2 on $\ell_o(Y) \equiv \{g \in \ell^2(Y),$ supp g finite $\}$. For Im $k > 0$ large enough one has $(\Gamma_{\alpha,Y}(k))^{-1}$ bounded. $\Gamma_{\alpha,Y}(k)$ is analytic in k for Im $k > 0$.

The proof exploits the monotonicity properties of the resolvent $(-\Delta_{\tilde{\alpha},\tilde{Y}} - k^2)^{-1}$ in $\tilde{Y}$, for k^2 sufficiently large.

<u>Remark:</u> It is possible to show that $-\Delta_{\alpha,Y}$ is local in the sense that if $\psi \in D(-\Delta_{\alpha,Y})$ and $\psi = 0$ in a domain U of $\mathbb{R}^3$, then $-\Delta_{\alpha,Y}\psi = 0$ in U. Moreover it is possible to approximate in norm resolvent sense $-\Delta_{\alpha,Y}$ by local scaled short range interactions, extending 1.1, a2): see [11].

1.4 Periodic point interactions

The case of "periodic point interactions" is of particular importance in solid state physics. In this case both α and Y are periodic, e.g. $Y = \mathbb{Z}^3$ and α constant. The corresponding Schrödinger operator $-\Delta_{\alpha,Y}$ of 1.3 is then the mathematical realization of the heuristic Hamiltonian $-\Delta + \lambda \sum_{y \in \mathbb{Z}^3} \delta_y(\cdot)$, after renormalization of the coupling constant λ (independent of y). This Hamiltonian fits in the so called "one-electron model of a solid", in as much as the solid is exemplified by a fixed infinitely extended crystal with (infinitely massive) nuclei at the vertices, and the electron moving in the crystal interacts only with the crystal (other effects like relativistic effects, lattice vibrations, spin-orbit coupling,..., are neglected, in this picture). More generally, we can consider heuristic point interaction Hamiltonians of the following form. Let Λ be a lattice of $\mathbb{R}^3$, i.e. $\Lambda \equiv \{\sum_{i=1}^{3} n_i a_i\}$, with a_1, a_2, a_3 a basis of $\mathbb{R}^3$ and $n \equiv (n_1, n_2, n_3)$ running over $\mathbb{Z}^3$.

Λ is called a Bravais lattice.

Let $\hat{\Gamma}$ be the basic period cell or primitive cell relative to Λ i.e. $\hat{\Gamma} = \mathbb{R}^3/\Lambda$ (so that each $x \in \mathbb{R}^3$ can be written as $x = \lambda + \hat{\gamma}$, with $\lambda \in \Lambda$ and $\hat{\gamma} \in \hat{\Gamma}$). $\hat{\Gamma}$ can be identified with the so called Wigner-Seitz cell $\{\sum_{i=1}^{3} s_i a_i\}$, $s_i \in [-\frac{1}{2}, \frac{1}{2})$.

Let y_j, $j = 1 \ldots N$ be N points in $\hat{\Gamma}$ i.e. $y_j \in \hat{\Gamma}$.

Then we can consider the heuristic Hamiltonian:

$$H = -\Delta + \sum_{\lambda \in \Lambda} \sum_j \lambda_j \, \delta_{y_j+\lambda}(\cdot). \tag{1.4}$$

This yields a model of a multiatomic crystal or a perfect alloy (think of $y_1 \ldots y_N$ as carrying N different atoms or nuclei, acting with coupling constants $\lambda_1,\ldots,\lambda_N$ by a δ-potential on an electron entering $\hat{\Gamma}$- and translate the whole picture by any $\lambda \in \Lambda$).

Of course for a proper mathematical definition of a self-adjoint operator associated with the above heuristic Hamiltonian we can use the theorem in 1.3, with $Y = \{y_j + \lambda,\ \lambda_j \in \hat{\Gamma},\ \lambda \in \Lambda\}$. Then the corresponding $-\Delta_{\alpha,Y}$ is the mathematical realization of H. For the detailed study of $-\Delta_{\alpha,Y}$, in particular its spectrum, we shall have to exploit the particular symmetry properties of (α,Y). It is useful to think of $-\Delta_{\alpha,Y}$ as coming from the interaction $\sum_{y \in \Lambda} \sum_j \delta_{y_j+\lambda}(\cdot)$ and reason on its invariance properties on the basis of those of the potential V, formally $\sum_{\lambda \in \Lambda} \sum_j \lambda_j \, \delta_{y_j+\lambda}(\cdot)$. More generally, it is useful to clarify the picture by studying Schrödinger operators of the form $-\Delta + V$ in $L^2(\mathbb{R}^3)$, with V the multiplication operator by a periodic, say smooth function with periods the vectors λ in a Bravais lattice Λ. By the smoothness and periodicity of V we can expand V in Fourier series i.e. $V(x) = \sum_{\gamma \in \Gamma} V_\gamma e^{i\gamma x}$, with uniform convergence, where

$$V_\gamma = |\Gamma|^{-1} \int_{\hat{\Gamma}} V(\nu)\, e^{-i\gamma\nu}\, d\nu. \tag{1.5}$$

Γ is the so called dual or reciprocal or orthogonal lattice i.e. $\Gamma = \{\sum_i n_i b_i\}$, with b_i, $i=1,2,3$ the dual basis of $\mathbb{R}^3$ given by $a_j b_{j'} = 2\pi\delta_{jj'}$, with $n = (n_1,n_2,n_3)$ running over $\mathbb{Z}^3$.

It is useful to consider also the Fourier transform $\hat{H}$ of H.

Let $\hat{V}(p) = (2\pi)^{-3/2} \int_{\mathbb{R}^3} V(x) e^{-ipx} dx$ be the Fourier transform of V (which exists if we look upon V e.g. as a tempered distribution; in fact, V being bounded, $\hat{V}$ is a pseudo measure).

The Fourier inversion formula holds i.e.

$$V(x) = (2\pi)^{-3/2} \int_{\mathbb{R}^3} \hat{V}(p)\, e^{ipx} dp, \tag{1.6}$$

in the sense of tempered distributions.

Let us now assume e.g. $V \in C_0^\infty(\mathbb{R}^3)$. Then the Fourier series expansion for V converges uniformly and one has

$$\hat{V}(p) = (2\pi)^{3/2} \sum_{\gamma \in \Gamma} V_\gamma \delta(p-\gamma), \tag{1.7}$$

with convergence e.g. in the weak topology on $\mathcal{S}'(\mathbb{R}^3)$.

Let $\mathfrak{F}$ be the unitary operator given by Fourier transform in $L^2(\mathbb{R}^3,dx)$. Then, as well known, $\mathfrak{F}(-\Delta)\mathfrak{F}^*$ is multiplication by p^2 in $L^2(\mathbb{R}^3,dp)$.
$\mathfrak{F} V \mathfrak{F}^*$ is the convolution operator in $L^2(\hat{\mathbb{R}}^3,dp)$:

$$(2\pi)^{-3/2}(\overline{\hat{V}}(p-\cdot),\cdot) = \sum_{\gamma \in \Gamma} V_\gamma \delta(p-\gamma-\cdot,\cdot). \tag{1.8}$$

Let U be the mapping from $L^2(\hat{\mathbb{R}}^3,dp)$ onto $\int_{\hat{\Lambda}}^{\oplus} \ell^2(\Gamma)d\theta \equiv L^2(\hat{\Lambda}, \ell^2(\Gamma))$ given by

$(U\hat{f})(\theta,\gamma) \equiv \hat{f}(\theta+\gamma)$, where $\hat{f} \in L^2(\hat{\mathbb{R}}^3,dp)$, with the identification $\hat{p} \leftrightarrow \theta+\gamma$, $\gamma \in \Gamma$, $\theta \in \hat{\Lambda}$,

with $\hat{\Lambda} \equiv \hat{\mathbb{R}}^3/\Gamma$ the dual group to Λ (also called basic periodic cell or primitive cell of the dual lattice Γ). $\hat{\Lambda}$ can be identified with the Wigner-Seitz cell $\{\sum_{i=1}^{3} s_i b_i \mid s_i \in [-\frac{1}{2},\frac{1}{2}),\ i=1,2,3\}$ of Γ, also called Brillouin zone.

Then $U\mathfrak{F}(-\Delta+V)\mathfrak{F}^* U^* = \int_{\hat{\Lambda}}^{\oplus} \hat{H}(\theta)d\theta$, with $\hat{H}(\theta)$ acting in $\ell^2(\Gamma)$ according to, as

easily deduced from (1.8):

$$(\hat{H}(\theta)g)(\gamma) = |\gamma+\theta|^2 g(\gamma) + \sum_{\gamma' \in \Gamma} V_{\gamma'} g(\gamma-\gamma'). \tag{1.9}$$

The sum on the r.h.s. converges for $g \in \ell^2(\Gamma)$.

$$\text{Thus } (U\mathfrak{F}(-\Delta+V)\mathfrak{F}^* U^* \hat{f})(\theta,\gamma) = (\hat{H}(\theta)\hat{f}(\theta,\cdot),\cdot)(\gamma),\ \forall f \in L^2(\hat{\mathbb{R}}^3,dp). \tag{1.10}$$

In particular the spectrum of $-\Delta+V$ is obtained by determining the spectrum of $\hat{H}(\theta)$. Now with $-\hat{\Delta} \equiv \mathfrak{F}(-\Delta)\mathfrak{F}^*$ we have

$$(-\hat{\Delta}(\theta)g)(\gamma) = |\gamma+\theta|^2 g(\gamma), \tag{1.11}$$

hence $\sigma(-\hat{\Delta}(\theta)) = \sigma_d(-\hat{\Delta}(\theta)) = |\Gamma+\theta|^2$, where σ_d means discrete spectrum and $|\Gamma+\theta^2| \equiv \{|\gamma+\theta|^2 \mid \gamma \in \Gamma\}$.

If e.g. $V_\gamma \in \ell^2$, then one shows, see e.g. [1], [3],

$$\sigma(H) = \sigma(\hat{H}) = \bigcup_{\theta \in \hat{\Lambda}} \sigma(\hat{H}(\theta)), \tag{1.12}$$

with $\hat{H} \equiv \mathfrak{F} H \mathfrak{F}^*$. $\sigma(\hat{H}(\theta))$ is purely discrete, consisting of isolated eigenvalues of finite multiplicity and $\sigma(\hat{H})$ is absolutely continuous and consists of bands separated by gaps.

How can one prove similar results in the case where V is a periodic point interaction, and hence is too singular to satisfy the above assumptions?

Two equivalent procedures can be followed. Either start with H replaced by $-\Delta_{\alpha,Y}$ and decompose according to

$$U(-\Delta_{\alpha,Y})U^* = \int_{\hat{\Lambda}}^{\oplus} (-\hat{\Delta}_{\alpha,Y})(\theta)d\theta. \tag{1.13}$$

Or use a perturbation theorem to perturb first $-\hat{\Delta}(\theta)$ to $-\hat{\Delta}(\theta)$ plus point interaction and integrate afterwards.

We shall describe here shortly the latter procedure. Formally we have to perturb $-\Delta(\theta)$ by $V(x) = -\sum_{j=1}^{N}\sum_{\lambda\in\Lambda}\mu_j\,\delta(x-y_j-\lambda)$ for some $\mu_j\in\mathbb{R}$ (which are going to be renormalized), y_i being points in the basic periodic cell $\hat{\Gamma}$. In this case $V_\gamma = -|\hat{\Gamma}|^{-1}\sum_{j=1}^{N}\mu_j\, e^{-i\gamma y_j}$, $\gamma\in\Gamma$, with $|\hat{\Gamma}|$ the volume of Γ.

Introduce for any $\kappa>0$ the operator in $\ell^2(\Gamma)$ (with cut off κ):

$$(\hat{H}^{\kappa}(\theta)g)(\gamma) \equiv |\gamma+\theta|^2\, g(\gamma) - |\hat{\Gamma}|^{-1}\sum_{j=1}^{N}\mu_j(\kappa)(\phi^{\kappa}_{y_j}(\theta),g)\,\phi^{\kappa}_{y_j}(\theta), \tag{1.14}$$

where $\theta\in\hat{\Lambda}$, $\gamma\in\Gamma$, $g\in\ell^2_o(\Gamma)$.

$(,)$ is the scalar product in $\ell^2(\Gamma)$.

$$\phi^{\kappa}_{y_j}(\theta)\,(\gamma) \equiv \chi_\kappa(\gamma+\theta)\, e^{-i(\gamma+\theta)y_j}, \tag{1.15}$$

with χ_κ the characteristic function of the closed ball in $\mathbb{R}^3$ with radius κ and center at the origin. We then have the

<u>Theorem 2</u>: Let $\hat{H}^{\kappa}(\theta)$ be the self-adjoint operator in $\ell^2(\Gamma)$ given by (1.14) with domain $D(\hat{H}^{\kappa}(\theta)) = D(-\hat{\Delta}(\theta)) = \{g\in\ell^2(\Gamma)\,|\,\sum_{\gamma\in\Gamma}|\gamma+\theta|^4\,|g(\gamma)|^2<\infty\}$, $\forall\,\theta\in\hat{\Lambda}$.

Then if $\mu_j(\kappa) = (\alpha_j+\kappa/2\pi^2)^{-1}$ for some $\alpha_j\in\mathbb{R}$, then $\hat{H}^{\kappa}(\theta)$ converges for all $\theta\in\hat{\Lambda}$ in norm resolvent sense as $\kappa\to\infty$ to a self-adjoint operator $-\hat{\Delta}_{\alpha,\Lambda,Y}(\theta)$, with $Y\equiv\{y_1,\dots,y_N\}$, $\alpha\equiv(\alpha_1\dots\alpha_N)$, whose resolvent is given by

$$(-\hat{\Delta}_{\alpha,\Lambda,Y}(\theta)-k^2)^{-1} = G'_k(\theta) + |\hat{\Gamma}|^{-1}\sum_{j,j'=1}^{N}[\Gamma_{\alpha,\Lambda,Y}(k,\theta)]^{-1}_{j,j'}\,(F_{-\bar{k},y_j}(\theta),\cdot)F_{k,y_j}(\theta), \tag{1.16}$$

for all k s.t. $k^2\notin|\Gamma+\theta|^2$, $\operatorname{Im}k>0$, $\det\Gamma_{\alpha,\Lambda,Y}(k,\theta)\neq 0$, $\theta\in\hat{\Lambda}$, where $\Gamma_{\alpha,\Lambda,Y}(k,\theta)$ is the $N\times N$ matrix with jj'-element $\alpha_j\,\delta_{jj'} - g_\kappa(y_j-y_{j'},\theta)$, with

$$g_\kappa(x,\theta) \equiv \begin{cases} |\hat{\Gamma}|^{-1}\lim\limits_{\kappa\to\infty}\sum\limits_{\substack{\gamma\in\Gamma\\|\gamma+\theta|\le\kappa}} e^{i(\gamma+\theta)x}\,/\,(|\gamma+\theta|^2-k^2) = \sum\limits_{\lambda\in\Lambda}G_k(x+\lambda)e^{-i\theta\lambda}, & x\in\mathbb{R}^3-\Lambda\\[2ex] (2\pi)^{-3}\,e^{-i\theta x}\lim\limits_{\kappa\to\infty}\Big[\sum\limits_{\substack{\gamma\in\Gamma\\|\gamma+\theta|\le\kappa}}(|\gamma+\theta|^2-k^2)^{-1}\,|\hat{\Lambda}| - 4\pi\kappa\Big] = \sum\limits_{\lambda\in\Lambda}\tilde{G}_k(x+\lambda)e^{-i\theta\lambda}+\dfrac{ik}{4\pi}, & x\in\Lambda, \end{cases} \tag{1.17}$$

$G'_k(\theta)$ the right hand side of (1.17), is the multiplication operator in $\ell^2(\Gamma)$ by the function $(|\gamma+\theta|^2-k^2)^{-1}$.

The term $G_k(x+\theta)$, $x\in\mathbb{R}^3-\Lambda$, $\theta\in\hat{\Lambda}$ appearing in the second expression for $g_k(x+\theta)$, is $(4\pi|\theta+x|)^{-1}e^{ik|\theta+x|}$, as in 1.2.

$\tilde{G}_k$ is by definition $\tilde{G}_k(y-y') = (1-\delta_{y-y',0})\, G_k(y-y') = 0$ for $y = y'$, $= G_k(y-y')$ for $y \neq y'$, $y,y' \in \mathbb{R}^3$.

Furthermore $F_{k,y_j}(\theta)(\gamma) \equiv (|\gamma+\theta|^2-k^2)^{-1}e^{-i(\gamma+\theta)y_j}$.

Proof (Sketch): From(1.14)we compute easily $(\hat{H}^{\kappa}(\theta)-k^2)^{-1}$.

Control for $\kappa\to\infty$ is by using the following Poisson formula for $G_k(\cdot)$, valid for all $a \in \mathbb{R}^3$, $\theta\in\Lambda$, Im $k>0$, and proven in [34], [11]:

$$\sum_{\substack{\lambda\in\Lambda\\ \lambda\neq -q}} (4\pi|\lambda+a|)^{-1}e^{ik|\lambda+a|}\,e^{-i\theta\lambda} = |\hat{\Gamma}|^{-1}\lim_{\kappa\to\infty}\sum_{\substack{\gamma\in\Gamma\\ |\gamma+\theta|\leq\kappa}} (|\gamma+\theta|^2-\kappa^2)^{-1}e^{i(\gamma+\theta)a} \text{ for}$$

$a\in\mathbb{R}^3-\Lambda$, $= (2\pi)^{-3}\,e^{i\theta a}\lim_{\kappa\to\infty}\Big[\sum_{\substack{\gamma\in\Gamma\\ |\gamma+\theta|\leq\kappa}} (|\gamma+\theta|^2-k^2)^{-1}\,|\hat{\Lambda}| - 4\pi\kappa\Big] - \frac{ik}{4\pi}$, for $a\in\Lambda$. ∎

The above computation of the operator $-\hat{\Delta}_{\alpha,\Lambda,Y}(\theta)$ is important inasmuch as this operator determines the Fourier transform $-\hat{\Delta}_{\alpha,Y+\Lambda} = \mathcal{F}(-\Delta_{\alpha,Y+\Lambda})\mathcal{F}^*$ of the Hamiltonian $-\Delta_{\alpha,Y+\Lambda}$ for point interactions at $Y+\Lambda$ with strengths $\alpha_j \equiv \alpha_{y_j+\lambda}$ (independent of $\lambda\in\Lambda$), $y_j\in Y\subset\hat{\Gamma}$, $j=1,\dots,N$.

In fact one proves

$$\int_{\hat{\Lambda}}^{\oplus}(-\hat{\Delta}_{\alpha,\Lambda,Y})(\theta)d\theta = U(-\hat{\Delta}_{\alpha,Y+\Lambda})U^*, \tag{1.18}$$

with U defined as in the paragraph after (1.8).

Moreover, by taking Fourier transforms:

$$\tilde{U}(-\Delta_{\alpha,Y+\Lambda})\tilde{U}^* = |\tilde{\Lambda}|^{-1}\int_{\hat{\Lambda}}^{\oplus}[-\Delta_{\alpha,\Lambda,Y}(\theta)]\,d\theta\ , \tag{1.19}$$

with $\tilde{U}$ the operator from $\mathcal{S}(\mathbb{R}^3)$ into $L^2(\hat{\Lambda}, |\hat{\Lambda}|^{-1}d\theta; L^2(\hat{\Gamma})) = |\hat{\Lambda}|^{-1}\int_{\hat{\Lambda}}^{\oplus}L^2(\hat{\Gamma})d\theta$, given by

$(\tilde{U}f)(\theta,\nu) \equiv \sum_{\lambda\in\Lambda} f(\nu+\lambda)\,e^{-i\theta\lambda}$, $\theta\in\hat{\Lambda}$, $\nu\in\hat{\Gamma}$, for all $f\in\mathcal{S}(\mathbb{R}^3)$.

Of course $-\hat{\Delta}_{\alpha,\Lambda,Y}(\theta) = \mathcal{F}(-\Delta_{\alpha,\Lambda,Y}(\theta))\mathcal{F}^*$ (1.20)

$-\Delta_{\alpha,\Lambda,Y}(\theta)$is self-adjoint acting in $L^2(\hat{\Gamma})$ with resolvent deduced from the one of $-\hat{\Delta}_{\alpha,\Lambda,Y}(\theta)$, which is (1.16) ,with G_k' replaced by g_k, $|\hat{\Gamma}|$ by $|\hat{\Lambda}|$, $F_{k,y_j}(\theta)$ by $g_k(\cdot-y_j,\theta)$.

We note that $g_k(\nu,\theta)$, $\nu\in\hat{\Gamma}$, $\theta\in\hat{\Lambda}$ is given by the formula (1.17) for $g_k(x,\theta)$ by simply replacing x by ν. The operator $g_k(\theta)$ is actually $(-\Delta(\theta)-k^2)^{-1}$ in $L^2(\hat{\Gamma})$, with $-\Delta(\theta)$ the self-adjoint operator $-\Delta$ on $L^2(\hat{\Gamma})$ with boundary conditions

$f(\nu+a_j) = e^{i\theta a_j}f(\nu)$, $\frac{\partial f}{\partial x_j}(\nu+a_j) = e^{i\theta a_j}\frac{\partial f}{\partial x_j}(\nu)$, $\theta\ \hat{\Lambda}$, with $\nu,\nu+a_j\in\partial\Gamma$, $j=1,2,3$.

Having these connections, the spectrum of the relevant Hamiltonian $-\Delta_{\alpha,Y+\Lambda}$ is determined by a detailed study of the spectrum of $-\hat{\Delta}_{\alpha,\Lambda,Y}(\theta)$, together with general results on spectra of direct integrals of operators. We formulate the spectral results, of direct relevance also in the stochastic case, first in the case where Y consists of just one point i.e. N = 1, which by translation by vectors of Λ can be assumed to be at the origin i.e. Y = {0}.

We shall simplify notations in this case and set $-\hat{\Delta}_{\alpha,\Lambda} \equiv \hat{\Delta}_{\alpha,\{o\}+\Lambda}$,

$$-\hat{\Delta}_{\alpha,\Lambda}(\theta) \equiv -\hat{\Delta}_{\alpha,\Lambda,\{o\}}(\theta) \tag{1.21}$$

Note that in this case α has only one value, associated with 0 (hence $\alpha = \alpha_o$ in the previous notation). The detailed structure of the spectrum of $-\hat{\Delta}_{\alpha,\Lambda}(\theta)$ is as follows, cfr. [34], [11] :

<u>Theorem 3:</u> $\sigma(-\hat{\Delta}_{\alpha,\Lambda}(\theta))$ is purely discrete and consists of isolated eigenvalues of finite multiplicity for all $\theta \in \hat{\Lambda}$, thus

$$\sigma_{ess}(-\hat{\Delta}_{\alpha,\Lambda}(\theta)) = \emptyset,\ \theta \in \hat{\Lambda}. \quad \mathbb{R} - |\Gamma + \theta|^2 = \bigcup_{n=0}^{\infty} I_n(\theta),$$

with $I_n(\theta)$ disjoint open intervals. In each $I_n(\theta)$, $-\hat{\Delta}_{\alpha,\Lambda}(\theta)$ has exactly one simple eigenvalue $E_n^{\alpha,\Lambda}(\theta)$ with eigenfunction

$$\psi_{E_n^{\alpha,\Lambda}(\theta)}(\gamma) = [|\gamma + k|^2 - E_n^{\alpha,\Lambda}(k)]^{-1},\ \gamma \in \Gamma.$$

$E^{\Lambda}(\theta) \in |\Gamma + \theta|^2$ is an eigenvalue of $-\hat{\Delta}_{\alpha,\Lambda}(\theta)$ of multiplicity $m \geq 1$ iff there exist $m + 1$ points $\gamma_o,\ldots,\gamma_m \in \Gamma$ such that $E^{\Lambda}(\theta) = |\gamma_o + \theta|^2 = \ldots = |\gamma_m + \theta|^2$.

The corresponding eigenspace is spanned by the eigenfunctions

$$\psi^{(j)}_{E^{\Lambda}(\theta)}(\gamma) = \delta_{\gamma\gamma_j} - \delta_{\gamma\gamma_o};\ \gamma,\ \gamma_j \in \Gamma,\ j = 1,\ldots,m.$$

There is a natural 1-1 correspondence $\Gamma \to \sigma(-\hat{\Delta}_{\alpha,\Lambda}(\theta)) = \{E_\gamma^{\alpha,\Lambda}(\theta)\}$, when $\sigma(-\hat{\Delta}_{\alpha,\Lambda}(\theta))$ is considered with multiplicities: if $|\gamma + \theta|^2$ for $\gamma \in \Gamma$ is no eigenvalue define $E_\gamma^{\alpha,\Lambda}(\theta)$ to be the largest eigenvalue smaller than $|\gamma + \theta|^2$. If $|\gamma + \theta|^2$ is an eigenvalue of multiplicity m there are m + 1 points $\gamma_o,\ldots,\gamma_m$ in Γ s.t. $|\gamma_o + \theta|^2 = |\gamma_1 + \theta|^2 = \ldots = |\gamma_m + \theta|^2$, and let $E_{\gamma_{j_o}}^{\alpha,\Lambda}(\theta)$ be as above for one $j_o \in \{0,\ldots,m\}$ and $E_{\gamma_j}^{\alpha,\Lambda}(\theta) = |\gamma + \theta|^2$ for all $j \neq j_o$. The lowest eigenvalue is $E_o^{\alpha,\Lambda}$ and is the solution with minimal k of $\alpha = g_k(0,\theta)$. No other eigenvalues of $-\hat{\Delta}_{\alpha,\Lambda}(\theta)$ exist.

<u>Proof:</u> (Sketch): The unperturbed $-\hat{\Delta}(\theta)$ (no point interactions) has purely discrete spectrum with eigenvalues $|\Gamma + \theta|^2$. From the explicit expression for the resolvent of $-\hat{\Delta}_{\alpha,\Lambda,Y}(\theta)$, in our case $-\hat{\Delta}_{\alpha,\Lambda}(\theta)$, we get that the eigenvalues k^2 of $-\hat{\Delta}_{\alpha,\Lambda}(\theta)$

are to found solely under the solutions of $\alpha = g_k(0,\theta)$ (which yields singularities of $(\Gamma_{\alpha,\Lambda}(k,\theta))^{-1}$ (with the notation $\Gamma_{\alpha,\Lambda} \equiv \Gamma_{\alpha,\Lambda,Y}$ for $Y = \{0\}$) or have to have the form $|\Gamma + \theta|^2$ (which yields a singularity of $G_k(\theta)$ and $F_{k,0}(\theta,\gamma)$).

One then observes that $g_k(0,\theta)$ increases in k^2 from the value $-\infty$ at $k^2 = -\infty$ and has poles at $|\Gamma+\theta|^2$. This yields the union structure of $\mathbb{R} - |\Gamma+\theta|^2$, and the unique eigenvalue $E_n^{\alpha,\Lambda}(\theta)$ on $I_n(\theta)$. Since $E^\Lambda(\theta) \equiv |\gamma+\theta|^2$ is a singularity in all terms of the resolvent, the study of the eigenvalues in $|\Gamma+\theta|^2$ is more delicate, see [11]. □

From the structure of the spectrum $-\hat{\Delta}_{\alpha,\Lambda}(\theta)$ one deduces then the one of $-\hat{\Delta}_{\alpha,\Lambda}$ and hence, by unitary invariance, of the Hamiltonian $-\Delta_{\alpha,\Lambda}$ with periodic point interactions on Λ. Let us set $\theta_o = -\frac{1}{2}(b_1+b_2+b_3)$, b_i being the basis of the dual lattice Γ.

<u>Theorem 4:</u> The spectrum $\sigma(-\Delta_{\alpha,\Lambda})$ is absolutely continuous, and is given by $\sigma(-\Delta_{\alpha,\Lambda}) = [E_o^{\alpha,\Lambda}(0), E_o^{\alpha,\Lambda}(\theta_o)] \cup [E_1^{\alpha,\Lambda}(\tilde{\theta}),\infty)$, where $E_o^{\alpha,\Lambda}(\theta) \leq E_1^{\alpha,\Lambda}(\theta)$ are the two smallest eigenvalues of $-\Delta_{\alpha,\Lambda}(\theta)$, and $E_o^{\alpha,\Lambda}(0)$ resp. $E_o^{\alpha,\Lambda}(\theta_o)$ are the minimum resp. maximum of $\{E_o^{\alpha,\Lambda}(\theta),\ \theta \in \hat{\Lambda}\}$ and $E_1^{\alpha,\Lambda}(\tilde{\theta}) \equiv \min_{\theta\in\hat{\Lambda}} E_1^{\alpha,\Lambda}(\theta) \geq \tilde{\theta}^2$. Let α_o be s.t. $E_o^{\alpha_o,\Lambda}(\theta_o) = 0$, then for all $\alpha \leq \alpha_o$ there is a gap in the spectrum. Moreover there exists an α_1 s.t. there is no gap in the spectrum for $\alpha \geq \alpha_1$.

<u>Proof:</u> See [34], [11], [42].

<u>Remark:</u> The presently available results on the detailed structure of the spectrum of $-\Delta_{\alpha,Y+\Lambda}$ in the case where Y consists of $N > 1$ points are less detailed. We only mention here that the negative part of $\sigma(-\Delta_{\alpha,\Lambda+Y})$ consists of at most N disjoint, closed intervals i.e. bands (see [11]).

<u>Remark.</u> The general approximation theorem permits to approximate $-\Delta_{\alpha,Y+\Lambda}$ in norm resolvent sense by scaled short range Hamiltonians

$H_{\varepsilon,Y+\Lambda} = -\Delta+\varepsilon^{-2}\sum_{j=1}^{N}\sum_{\lambda\in Y}\lambda_j(\varepsilon)V_j((\cdot-y_j-\lambda/\varepsilon)$, with $\lambda_j \in C^2$ at $\varepsilon=0$. By decomposing $H_{\varepsilon,Y+\Lambda}$ as $|\Lambda|^{-1}\int_\Lambda^\oplus H_{\varepsilon,\Lambda,Y}(\theta)d\theta$ and studying the spectrum of the "reduced" Hamiltonian $H_{\varepsilon,\Lambda,Y}(\theta)$, one arrives at approximation results for the spectrum of $-\Delta_{\alpha,Y+\Lambda}$. In fact $H_{\varepsilon,\Lambda,Y}(\theta)$ converges in norm resolvent sense to $-\Delta_{\alpha,\Lambda,Y}$ with suitable α determined by the λ_j at 0 (similarly as in Sect. 1.1). In particular for $Y = \{0\}$ one shows that the bands connected at points $E(\hat{\theta}_o)$ do not open up to first order in ε. For details see [11].

2. Random point interactions in 3-dimensions

In this section we shall study Schrödinger Hamiltonians with point interactions, $-\Delta_{\alpha,Y}$, where both the location of sources Y and the strengths are random. After defining the operators we shall study their spectrum and compare it with the one of the deterministic counterpart.

2.1 Definition of the random Hamiltonian and its spectrum

Let $(\Omega, \mathbf{F}, P)$ be a probability space and let $\{Y(\omega), \omega \in \Omega\}$ be a countable random subset of $\mathbb{R}^3$ of the form $Y(\omega) = \{y_j(\omega), j \in \mathbb{N}\}$, where the y_j are $\mathbb{R}^3$-valued random variables s.t. $Y(\omega)$ has no accumulation points i.e. $\inf_{j \neq j' \in \mathbb{N}} |y_j(\omega) - y_{j'}(\omega)| > 0 \ \forall \omega \in \Omega$. Let $\alpha(\omega) \equiv \{\alpha_{y_j}(\omega), j \in \mathbb{N}\}$ be an Y-indexed family of real-valued random variables, with $\alpha_{y_i}(\omega) \in \mathbb{R}, \ \forall y_j \in Y \ \ \forall \omega \in \Omega$.

Then $H_\omega \equiv -\Delta_{\alpha(\omega),Y(\omega)}$ is a well defined self-adjoint operator in $L^2(\mathbb{R}^3)$. Its resolvent is given by (1.2) with a replaced by $\alpha(\omega)$ and Y replaced by $Y(\omega)$. The domain properties and the locality property of H_ω hold as for the corresponding operator $-\Delta_{\alpha,Y}$, as well as of course the approximation by local scaled short range random interactions $H_{\varepsilon,Y(\omega)}$. We are particularly interested in the study of H_ω and its spectrum when $\alpha(\omega)$, $Y(\omega)$ have some ergodicity property so that $\sigma(H_\omega)$ is indeed independent of ω, P-almost surely. We shall take up here the particularly simple case, where $Y(\omega)$ is the level 1 stochastic set of a countable family of independent identically distributed $\{0,1\}$-valued random variables i.e. $Y(\omega) = \{i \in \mathbb{Z}^3 | X_i(\omega) = 1\}$, looking upon $Y(\omega)$ as the "occupied sites" in the lattice $\mathbb{Z}^3$.

We shall assume $\alpha_j(\omega)$, $j \in \mathbb{Z}^3$ to be independent identically distributed random variables taking values in a fixed finite interval $[a,b]$ of $\mathbb{R}$: $\{X_j, j \in \mathbb{Z}^3\}$ and $\{\alpha_j, j \in \mathbb{Z}^3\}$ are $\mathbb{Z}^3$-indexed homogeneous random fields with values in $\{0,1\}$ resp. $[a,b]$.

We can think of (Ω, F, P) as the canonical space for the joint field $\{X_j(\omega), \alpha_j(\omega), j \in \mathbb{Z}^3\}$, so that $\{X_j(\omega), \alpha_j(\omega)\}$ is the j-th coordinate of $\omega \in \Omega$. Let T_j, $j \in \mathbb{Z}^3$ be the shift operator of Ω defined by $(T_j\omega)_{j'} \equiv \omega_{j'-j}$, $j,j' \in \mathbb{Z}^3$.

T_j is then a measurable transformation which preserves P. Let A be a T_j-invariant set i.e. $T_j(A) = A \ \forall j \in \mathbb{Z}^3$. Then A is in the tail σ-algebra of the random variables X_j, α_j. The X_j, α_j being independent we have $P(A) \in \{0,1\}$, by Kolmogorov 0-1 law. Hence T_j is P-ergodic. Let U_i, $i \in \mathbb{Z}^3$ be the family of unitary maps in $L^2(\mathbb{R}^3)$ expressing translation by i i.e. $(U_i f)(x) \equiv f(x-i)$, $\forall f \in L^2(\mathbb{R}^3)$, $x \in \mathbb{R}^3$.

For all $\omega \in \Omega$, $i \in \mathbb{Z}^3$, $k^2 \notin \sigma(H_\omega)$ we have

$$U_i (H_\omega - k^2)^{-1} U_i^* = (H_{T_i\omega} - k^2)^{-1}, \tag{2.1}$$

as seen by approximating H_ω in the norm convergent sense by short range interactions $H_{\varepsilon,Y(\omega)}$, with $\lambda_j(\varepsilon, T_i\omega) = \lambda_{j-i}(\varepsilon,\omega)$.

By a general result proven by Kirsch and Martinelli (cfr. e.g. [43]) we have then

Theorem 5: $\sigma(H_\omega) = \Sigma$, independent of ω, P-a.s.

Moreover $\sigma_{pp}(H_\omega) \equiv$ {closure of all eigenvalues of H_ω}, $\sigma_{sc}(H_\omega)$, $\sigma_c(H_\omega)$, $\sigma_{ac}(H_\omega)$ are also non random sets and the discrete part of $\sigma(H_\omega)$ is P-almost surely empty.

Proof: (Sketch, for details see e.g. [6], [43]). Let $\{E_\omega(\lambda)\}$ be the spectral family associated with H_ω, $f_{\lambda,\mu}(\omega) \equiv \dim \text{range} \{E_\omega(\lambda) - E_\omega(\mu)\} = \text{tr}[E_\omega(\lambda) - E_\omega(\mu)]$.

Step 1: $f_{\lambda,\mu}(\omega)$ is measurable and T_i-invariant, hence, by ergodicity, constant. I.e. there exists $\Omega_{\lambda,\mu} \subset \Omega$ s.t. $P(\Omega_{\lambda,\mu}) = 1$ and $f_{\lambda,\mu}(\omega) = C_{\lambda,\mu}$ $\forall \omega \in \Omega_{\lambda,\mu}$, with $C_{\lambda,\mu}$ independent of ω.

Then $\Omega_o \equiv \bigcap_{\lambda,\mu \in \mathbb{Q}} \Omega_{\lambda,\mu}$ satisfies $P(\Omega_o) = 1$ and $f_{\lambda,\mu}(\omega) = C_{\lambda,\mu}$ $\forall \omega \in \Omega_o$ $\forall\, \lambda,\mu \in \mathbb{Q}$.

The measurability and invariance of $f_{\lambda,\mu}$ are proven by using the relation $E_{T_i\omega}(\lambda) = U_i\, E_\omega(\lambda)\, U_i^*$, which follows from (2.1),and the unitary invariance of the trace.

Step 2: One uses the criterium: $\lambda \in \sigma(H_\omega) \Leftrightarrow \forall\, \varepsilon > 0$, $E_\omega(\lambda+\varepsilon) - E_\omega(\lambda-\varepsilon) \neq 0$, and the relation between $E_\omega(\lambda)$ and $f_{\lambda,\mu}$ and the already proven ω-independence of $f_{\lambda,\mu}$ (Step 1).

The invariance of $\sigma_{ess}(H_\omega)$ follows then using $\lambda \in \sigma_{ess}(H_\omega) \Leftrightarrow \forall\, \varepsilon > 0$,

$\dim \mathrm{Tr}(E_\omega(\lambda+\varepsilon) - E_\omega(\lambda-\varepsilon)) = +\infty$ and Step 1.

The invariance of other parts of the spectrum is more delicate, an essential ingredient being the measurability of the corresponding projection operators, e.g. $P_{ac}(\omega)$, the projection operator onto the absolutely continuous subspace of $L^2(\mathbb{R}^3)$. See [6], [43] for details. We shall now study the properties of the spectrum of H_ω.

2.2 Gaps in the spectrum of the stochastic Schrödinger operator

Let, as above, $X_i(\omega)$, $\alpha_i(\omega)$, $i \in \mathbb{Z}^3$ be identically distributed independent random variables, with values in $\{0,1\}$ resp. $[a,b]$. Let us denote by $\Phi(\omega)$ the family $\{X_i(\omega), \alpha_i(\omega), i \in \mathbb{Z}^3\}$. $\Phi(\omega)$ determines the positions and strengths of the random potential sources, hence we call $\Phi(\omega)$ the "stochastic potential". We denote by $H(\Phi(\omega))$ the corresponding Schrödinger operator $-\Delta_{\alpha(\omega),Y(\omega)}$ with point interactions given by $\Phi(\omega)$ i.e. $\alpha(\omega) \equiv \{\alpha_i(\omega), i \in \mathbb{Z}^3\}$, $Y(\omega) \equiv \{i \in \mathbb{Z}^3 | X_i(\omega) = 1\}$. A sequence $\varphi \equiv \{(\mu_i,\lambda_i), i \in \mathbb{Z}^3\}$ of values of $\Phi(\omega)$, with $\mu_i \in \{0,1\}$, $\lambda_i \in [a,b]$, is called an admissible potential. Let $\mathcal{A}$ be the class of all admissible potentials. For each $\varphi \in \mathcal{A}$ we denote by $H(\varphi)$ the Hamiltonian $-\Delta_{\alpha,Y}$ with $\alpha \equiv (\lambda_i, i \in \mathbb{Z}^3)$, $Y = \{i \in \mathbb{Z}^3 | \mu_i = 1\} \equiv Y(\varphi)$.

We call a $\varphi \in \mathcal{A}$ periodic with periods L_1, L_2, L_3 if there exist L_1, L_2, L_3 linear independent and belonging to the lattice Λ s.t. $\mu_{i+L_j} = \mu_i$, $\lambda_{i+L_j} = \lambda_i$ $\forall\, i \in \mathbb{Z}^3$, $j=1,2,3$. We denote by $\mathcal{P}$ the set of all admissible periodic potentials. The following theorem is proven in [6], [20] , see also e.g. [11] :

Theorem 6: Let $\Phi(\omega)$ be a stochastic potential. Then for any $\varphi \in \mathcal{A}$ we have:

i) $\sigma(H(\varphi)) \subset \Sigma (\equiv \sigma(H(\Phi(\omega)))$ P.a.s.)

ii) $\Sigma = \bigcup_{\varphi \in \mathcal{A}} \sigma(H(\varphi))$

iii) $\Sigma = \overline{\bigcup_{\varphi \in \mathcal{S}} \sigma(H(\varphi))}$,

where - means closure.

Proof: Let $\Phi(\omega)$ be the given stochastic potential and let $\varphi = \{(\mu_i, \lambda_i),\ i \in \mathbb{Z}^3\} \in \mathcal{A}$. Then, as easily seen from (1.3):

$$(H(\varphi)-k^2)^{-1}(x,y) = G_k(x-y) + \sum_{j,j' \in \mathbb{Z}^3} \{(\tilde{T}_k(\varphi))^{-1}(j,j') - \mathbb{1}_{(Y(\varphi))^c}(j,j')\, G_k(x-j) G_k(y-j'), \tag{2.2}$$

with $\mathbb{1}_A$ the identity on A,

$\tilde{T}_k(\varphi) \equiv T_k(\varphi) + \mathbb{1}_{(Y(\varphi))^c}$, $(Y(\varphi))^c \equiv \mathbb{Z}^3 - Y(\varphi)$ (the complement of $Y(\varphi)$), and

$T_k(\varphi) \equiv \Gamma_{\lambda, Y(\varphi)}$ i.e.

$$T_k(\varphi)(j,j') = (\lambda_j - \frac{ik}{4\pi})\delta_{jj'} - \tilde{G}_k(j-j'),\ j,j' \in Y(\varphi). \tag{2.3}$$

Define $\Omega^1 \equiv \{\omega \in \Omega \mid \sigma(H(\Phi(\omega))) = \Sigma\}$.
By Theor. 5 we have that Ω^1 has probability 1 i.e. $P(\Omega^1) = 1$.
Define for each $n \in \mathbb{N}$

$\Omega_n^{\varphi} \equiv \{\omega \in \Omega \mid\ |\alpha_i(\omega) - \lambda_i| < 1/n\ \forall\ |i| \leq n\}$.

The α_i being independent identically distributed we have:

$$P(\Omega_n^{\varphi}) = \prod_{|i| \leq n} P\{\omega \in \Omega \mid\ |\alpha_i(\omega) - \lambda_i| < 1/n\} = \prod_{|i| \leq n} P\{\omega \in \Omega \mid\ |\alpha_o(\omega) - \lambda_i| < 1/n\}.$$

But $\lambda_i \in \operatorname{supp} P_{\alpha_o} = [a,b]$. If we had $P\{\omega \in \Omega \mid\ |\alpha_o(\omega) - \lambda_i| < 1/n\} = 0$ for some $n \in \mathbb{N}$ then we would have

$P_{\alpha_o}((\lambda_i - \frac{1}{n}, \lambda_i + \frac{1}{n})) = 0$, hence $\lambda_i \notin \operatorname{supp} \alpha_o$. Thus $P\{|\alpha_o - \lambda_i| < 1/n\} > 0$, hence $P(\Omega_n^{\varphi}) > 0$. In particular then $\Omega^1 \cap \Omega_n^{\varphi} \neq \emptyset$ (otherwise $P(\Omega^1 \cup \Omega_n^{\varphi}) = 1 + P(\Omega_n^{\varphi}) > 1$!)

Let us take some $\omega_n \in \Omega^1 \cap \Omega_n^{\varphi}$. Since $\omega_n \in \Omega^1$ we have by the definition of Ω^1, $H(\Phi(\omega_n)) = \Sigma$. An easy computation, using the definition of Ω_n^{φ}, yields that (with ω_n for $\Phi(\omega_n)$):

$\tilde{T}_k(\omega_n)^{-1} - \mathbb{1}_{Y(\omega_n)^c} \to \tilde{T}_k(\varphi)^{-1} - \mathbb{1}_{Y(\varphi)^c}$ as $n \to \infty$, strongly on vectors in $\ell^2(\mathbb{Z}^3)$ of finite support. This then implies by the formula (2.2) and the corresponding one for $H(\Phi(\omega_n))$ that $H(\Phi(\omega_n)) \to H(\varphi)$ in the strong resolvent sense. This then implies

$$\sigma(H(\varphi)) \subset \sigma(H(\Phi(\omega_n))) = \Sigma \tag{2.4}$$

This proves i).
In particular then

$$\bigcup_{\varphi \in \mathcal{A}} \sigma(H(\varphi)) \subset \Sigma \tag{2.5}$$

Let $\omega^1 \in \Omega^1$, then $\sigma(H(\Phi(\omega^1))) = \Sigma$. (2.6)

Let $\varphi_1 \equiv \{(\mu_i,\lambda_i),\ i \in \mathbb{Z}^3\} = \Phi(\omega^1)$, with $\mu_i = X_i(\omega^1)$, $\lambda_i = \alpha_i(\omega^1)$.

Then $\varphi_1 \in \mathcal{A}$. Thus

$$\bigcup_{\varphi \in \mathcal{A}} \sigma(H(\varphi)) \supset \sigma(H(\varphi_1)) = \sigma(H(\Phi(\omega^1))). \tag{2.7}$$

From (2.6) and (2.7) we deduce

$$\bigcup_{\varphi \in \mathcal{A}} \sigma(H(\varphi)) \supset \Sigma . \tag{2.8}$$

From (2.8) and (2.5) we get

$$\bigcup_{\varphi \in \mathcal{A}} \sigma(H(\varphi)) = \Sigma,$$

which proves ii).

To prove iii) it suffices for any $\varphi \in \mathcal{A}$ to find $\varphi_n(\varphi) \in \mathcal{P}$ s.t. $H(\varphi_n(\varphi)) \to H(\varphi)$ strongly as $n \to \infty$. In fact then $\sigma(H(\varphi)) \subset \overline{\bigcup_n \sigma(H(\varphi_n(\varphi)))}$, with together with ii) yields

$$\Sigma \subset \overline{\bigcup_{\varphi \in \mathcal{P}} \sigma(H(\varphi))}. \tag{2.9}$$

But, on the other hand, since $\mathcal{P} \subset \mathcal{A}$ we have, by (2.4) $\Sigma \supset \bigcup_{\varphi \in \mathcal{P}} \sigma(H(\varphi))$.

This and (2.9) imply, by taking closures, iii).

That to given $\varphi \in \mathcal{A}$ one can find $\varphi_n(\varphi) \in \mathcal{P}$ s.t. $H(\varphi_n(\varphi)) \to H(\varphi)$ strongly as $n \to \infty$ is easily seen, as follows. Define in fact

$(\varphi_n(\varphi))_i \equiv (\mu_i^n(\varphi),\ \lambda_i^n(\varphi)) \equiv (\mu_i,\lambda_i)$ for $|i| \leq n$, $(\varphi_n(\varphi))_{i+(2n+1)\lambda} \equiv (\varphi_n(\varphi))_i$ for all $\lambda \in \Lambda$.

Take $\psi \in \ell_o^2 = \{\chi \in \ell^2(\mathbb{Z}^3) | \chi$ has finite support$\}$. Then

$\|[T_k(\varphi_n(\varphi)) - T_k(\varphi)]\psi\|^2_{\ell^2} = \sum_j |\lambda_j^n(\varphi) - \lambda_j|^2 |\psi(j)|^2 \to 0$ as $n \to \infty$, by the fact

that $\lambda_j^n(\varphi) = \lambda_j$ for $|j| \leq n$ and ψ has finite support. This implies the strong convergence of $T_k(\varphi_n(\varphi))$ to $T_k(\varphi)$ as $n \to \infty$. By the fact that $T_k(\varphi_n(\varphi))$ and $T_k(\varphi)$ are bounded and symmetric, this implies $T_k(\varphi_n(\varphi)) \to T_k(\varphi)$ in the strong resolvent sense. From (2.2) we then have $H(\varphi_n(\varphi)) \to H(\varphi)$ as $n \to \infty$, as claimed above. This completes the proof. □

We shall now study the behavior of the negative parts of the spectrum of $H(\Phi(\omega))$ as one removes point interactions.

Lemma 7. Let Λ_n be a sequence of cubic boxes in $\mathbb{Z}^3$, with size n, centered at the origin. Let $\varphi_{\alpha,n}$ the potential $\{(\mu_i,\alpha_i)\}$, with $\mu_i = 1$ for $i \in \Lambda_n$, $\mu_i = 0$ for $i \notin \Lambda_n$.

Then

$$\sigma(H(\varphi_{\alpha,n})) \cap (-\infty,0) \subset [E_o(a), E_1(b)] \cap (-\infty,0),$$

with $E_o(a) = E_o^{a,\mathbb{Z}^3}(0)$, $E_1(b) = E_o^{b,\mathbb{Z}^3}(\theta_o)$, in the notation of Theor. 4, the upper indices referring to the deterministic systems of periodic point interactions $\alpha_i = a\ \forall\ i \in \mathbb{Z}^3$ resp. $\alpha_i = b\ \forall i \in \mathbb{Z}^3$.

Proof: As we saw in the discussion of deterministic point interactions in Sect. 1.3, the negative eigenvalues of $H(\varphi_{\alpha,n})$ are given by the zero eigenvalues of $T_k(\varphi_{\alpha,n})$, defined by (2.3) with φ replaced by $\varphi_{\alpha,n}$. Let us denote by $k_{i,n}$ with $k_{i,n}^2 < 0$ these values. Let $E_{max,n}(\alpha)$ resp. $E_{min,n}(\alpha)$ be the largest and the smallest of the $E_{i,n}$ respectively. Due to the fact that $T_k(\varphi_{\alpha,n})$ is monotonically decreasing in $k^2 < 0$ we have that $E_{min,n}(\alpha)$ is the unique value of $k^2 < 0$ s.t. $\inf \sigma(T_k(\varphi_{\alpha,n})) = 0$. Similarly $E_{max,n}(\alpha)$ is the unique value of $k^2 < 0$ s.t. $\sup \sigma(T_k(\varphi_{\alpha,n})) = 0$. Moreover, by monotonicity in the α's:

$$\inf \sigma(T_k(\varphi_a)) \leq \inf \sigma(T_k(\varphi_{\alpha,n})),$$

$$\sup \sigma(T_k(\varphi_b)) \geq \sup \sigma(T_k(\varphi_{\alpha,n})).$$

Let $E_o(a)$ be the value of k^2 for which $\inf\sigma(T_k(\varphi_a)) = 0$ and let $E_1(b)$ the value of k^2 s.t. $\sup \sigma(T_k(\varphi_b)) = 0$. One easily shows, in the notation of Theor. 4, that

$E_o(a) = E_o^{a,\mathbb{Z}^3}(0)$, $E_1(b) = E_o^{b,\mathbb{Z}^3}(\theta_o)$.

From the above then we deduce $E_o(a) \leq E_{min,n}(\alpha) \leq E_{max,n}(\alpha) \leq E_1(b)$.

From this and Theor. 4 we deduce

$$\sigma(H(\varphi_{\alpha,n})) \cap (-\infty,0) \subset [E_o(a), E_1(b)] \cap (-\infty,0). \qquad \square$$

Lemma 8. Let Λ_n be as in Lemma 7. Let X be an arbitrary subset of $\mathbb{Z}^3$. Let $\varphi_{\alpha,n}^X$ be the potential $\{(x_i,\alpha_i)\}$ with $x_i = 1$ for $i \in \Lambda_n - X$, $x_i = 0$ for $i \notin \Lambda_n - X$. Then

$$\sigma(H(\varphi_{\alpha,n}^X)) \subset \sigma(H(\varphi_{\alpha,n})).$$

Proof. The matrix $T_k(\varphi_{\alpha,n}^X)$ is the restriction of $T_k(\varphi_{\alpha,n})$ to a subspace. It follows by a theorem of Rayleigh (see e.g. [3]) that

$$\inf \sigma(T_k(\varphi_{\alpha,n})) \leq \inf \sigma(T_k(\varphi_{\alpha,n}^X))$$

and (2.10)

$$\sup \sigma(T_k(\varphi_{\alpha,n}^X)) \leq \sup \sigma(T_k(\varphi_{\alpha,n})).$$

Define $E^X_{min,n}(\alpha)$, $E^X_{max,n}(\alpha)$ in the same way as we defined $E_{min,n}(\alpha)$ resp. $E_{max,n}(\alpha)$, replacing $\varphi_{\alpha,n}$ by $\varphi^X_{\alpha,n}$. The above inequality (2.10), together with the property that $T_k(\varphi^X_{\alpha,n})$ and $T_k(\varphi_{\alpha,n})$ are monotonically decreasing in $k^2 < 0$, yields

$$E_{min,n}(\alpha) \leq E^X_{min,n}(\alpha) \leq E^X_{max,n}(\alpha) \leq E_{max,n}(\alpha) \qquad (2.11)$$

This yields $\sigma(H(\varphi^X_{\alpha,n})) \cap (-\infty,0) \subset \sigma(H(\varphi_{\alpha,n})) \cap (-\infty,0)$.

But $\mathbb{R}_+$ belongs to $\sigma(H(\varphi^X_{\alpha,n}))$ as well as to $\sigma(H(\varphi_{\alpha,n}))$, since the potential vanishes outside the bounded region Λ_n. This proves the Lemma. □

Now let α in $\varphi_{\alpha,n}$, $\varphi^X_{\alpha,n}$ be periodic. Reasoning similarly as the proof of Theorem 6 we see that $H(\varphi_{\alpha,n}) \to H(\varphi_\alpha)$ and $H(\varphi^X_{\alpha,n}) \to H(\varphi^X_\alpha)$ in the strong resolvent sense, as $n \to \infty$, where

$$\varphi_\alpha = \{(\mu_i,\alpha_i)\} \qquad \text{with } \mu_i = 1 \quad \forall\, i \in \mathbb{Z}^3,$$
$$\varphi^X_\alpha = \{(\mu_i,\alpha_i)\} \qquad \text{with } \mu_i = 1 \quad \forall\, i \in \mathbb{Z}^3 - X,\ \mu_i = 0 \ \forall\, i \in X.$$

By Lemma 7, 8 we have then

$$\sigma(H(\varphi^X_\alpha)) \cap(-\infty,0) \subset \sigma(H(\varphi_\alpha)) \cap(-\infty,0) \subset [E_o(a),E_1(b)] \cap (-\infty,0) \qquad (2.12)$$

Now let $\Phi(\omega) = (\alpha(\omega), Y(\omega))$ be a stochastic potential. For some $\omega_1 \in \Omega$ we have

$$\sigma(H(\Phi(\omega_1)) \doteq \Sigma \qquad (2.13)$$

by Theor. 5. Let us take $X = (Y(\omega_1))^c$, $\alpha_i = (\alpha(\omega_1))_i \quad \forall\, i \in \mathbb{Z}^3$.

Then $\quad \Phi(\omega_1) = \varphi^X_\alpha. \qquad (2.14)$

From (2.12), (2.14) we get

$$\Sigma \cap (-\infty,0) \subset [E_o(a),\ E_1(b)] \cap (-\infty,0).$$

In particular

$$[E_1(b),0) \cap \Sigma = \emptyset.$$

We are ready to give a precise description of Σ. Let us denote by $P_{\alpha_o} \equiv P_o$ the common distribution of the α_i and let us set $q = P(X_i = 0)$. First, we consider the case where $q = 0$, i.e. "all lattice positions are occupied".

Theorem 9: Suppose that $q = P(X_i = 0) = 0$ and that

$$\inf(\operatorname{supp} P_o) = a,\ \sup(\operatorname{supp} P_o) = b$$

(i) If $E_o(b) \leq E_1(a)$ then $\Sigma = [E_o(a),\ E_1(b)] \cup [E_2,\infty) = \sigma(H_a) \cup \sigma(H_b)$, $E_2 \equiv E_1^{a,\Lambda}(\tilde{\theta})$

(ii) If $\operatorname{supp} P_o = [a,b]$

then $\Sigma = [E_o(a),\ E_1(b)] \cup [E_2,\infty) = \bigcup_{\gamma\in[a,b]} \sigma(H_\gamma)$

Proof: (i) and (ii) " $\subset$ ". Let ϕ be a periodic admissible potential. Then, by the previous discussion we have $\sigma(H(\phi)) \subset [E_o(a), E_1(b)] \cup [E_2,\infty)$.

Hence by Theorem 6 we have $\sum \subset [E_o(a), E_1(b)] \cup [E_2,\infty)$

(i) " $\supset$ " Conversely, if $\lambda \in [E_o(a), E_1(b)] \cup [E_2,\infty)$ then $\lambda \in \sigma(H_a)$ or $\lambda \in \sigma(H_b)$, since both H_a and H_b "have" admissible potentials (namely $\alpha_i \equiv a$ and $\alpha_i \equiv b$ resp.) we conclude $\lambda \in \sum$.

(ii) " $\supset$ " In the case (ii) we have by continuity that if $\lambda \in [E_o(a), E_1(b)] \cup [E_2,\infty)$ then $\lambda \in \sigma(H_\gamma)$ for some $\gamma \in [a,b]$. Again we conclude that $\lambda \in \Sigma$. □

Let us now suppose that $0 < q < 1$ (q as in Theor. 9). (The case $q = 1$ corresponds to the free Hamiltonian). In this case Theorem 9 remains valid as far as the negative spectrum is concerned. However, the whole half line $[0,\infty)$ belongs to the spectrum Σ as can be seen as follows. Since $P(X_i = 0) > 0$ the event that $X_i = 0$ for all $|i| \leq N$ has positive probability no matter how large N is. It follows that

$\Omega_N = \{\omega|$ there exists $i_o \in \mathbb{Z}^3$ such that $X_i = 0$ for $|i-i_o| \leq N\}$

has probability one, since Ω_N is invariant under translations (= shifts). Setting $\Omega_\infty = \bigcap_{N \in \mathbb{N}} \Omega_N$ we see that $P(\Omega_\infty) = 1$. But if $\omega \in \Omega_\infty$ we find boxes (perhaps far away) of arbitrary large size where the potential vanishes. Hence an easy argument using Weyl trial functions implies that $[0,\infty) \subset \sigma(H_\omega)$. So we have, using also Lemmas 7,8:

Theorem 10: Assume $0 < q < 1$

(i) If $a,b \in \operatorname{supp} P_o$, $\operatorname{supp} P_o \subset [a,b]$ and $E_o(b) \leq E_1(a)$ then

$$\Sigma = [E_o(a),E_1(b)] \cup [0,\infty) = \sigma(H_a) \cup \sigma(H_b) \cup \sigma(-\Delta)$$

(ii) If $\operatorname{supp} P_o = [a,b]$ then

$$\Sigma = [E_o(a),E_1(b)] \cup [0,\infty) = \bigcup_{\gamma \in [a,b]} \sigma(H_\gamma) \cup \sigma(-\Delta)$$

Theorems 9 and 10 can be looked upon as generalizations of the Saxon-Hutner conjecture (see [6], [20], [21], [7] and references therein).

There are numerous generalizations of the last two theorems. For example the condition of independence can be considerably weakened. Denote by P the distribution of the random field $\{\alpha_i, X_i\}$ realized on the probability space $[a,b]^{\mathbb{Z}^3} \times \{0,1\}^{\mathbb{Z}^3}$. It suffices for Theorem 9 to hold that supp P contains the sequences α_i with $\alpha_i \equiv a$ and $\alpha_i \equiv b$

for (i) and $\alpha_i \equiv \gamma$ for all $\gamma \in [a,b]$ for (ii). For Theorem 10 we have to assume that in addition to $\{\alpha_i, X_i\}$, α_i as above and $X_i \equiv 1$, also $\{\alpha_i, X_i\}$ with some α_i and $X_i \equiv 0$ belongs to supp P (see [46]). Moreover one may allow other types of randomness (e.g. random deviations from lattice positions).

In the next section we shall examine the case d = 1 which presents interesting new features as compared to the case d = 3.

3 Point interaction in one dimension

The definition of δ-interactions in one dimension is easier than in 3-dimensions since no renormalization of the coupling constant is needed. Let us start with the case of a one center point interaction, centered at the origin.

A. One point interaction

Let $\dot{H} = -\frac{d^2}{dx^2}$ on the domain $D(\dot{H}) = \{g \in H^{2,2}(\mathbb{R}) | g(0) = 0\}$,

where $H^{2,2}$ is the Sobolev space of generalized functions with square integrable derivatives up to second order. By the general theory of ordinary differential operators its adjoint is $(\dot{H})^* = -\frac{d^2}{dx^2}$ with domain $D(\dot{H})^*) = H^{2,2}(\mathbb{R} - \{0\}) \cap H^{2,1}(\mathbb{R})$.

$\dot{H}$ has deficiency indices (1,1). The self-adjoint extensions of $\dot{H}$ are given by the one parameter family $-\Delta_\alpha$ defined by $-\frac{d^2}{dx^2}$ with domain

$D(-\Delta_\alpha) = \{g \in H^{2,1}(\mathbb{R}) \cap H^{2,2}(\mathbb{R} - \{0\}) | \, g'(0+) - g'(0-) = \alpha g(0)\}$, for $\alpha \in \mathbb{R} \cup \{+\infty\}$ (and $g(0\pm) = \lim_{\varepsilon \downarrow 0} g(\pm\varepsilon)$). For $\alpha = 0$ we have $-\Delta_\alpha \equiv -\Delta = -\frac{d^2}{dx^2}$ on $D(-\Delta_0) = H^{2,2}(\mathbb{R})$. For $\alpha = +\infty$ we have $-\Delta_\infty = (-\Delta_{D_-}) \oplus (-\Delta)_{D_+}$, with $-\Delta_{D_\pm}$ the Dirichlet Laplacian on $(0, \pm\infty)$ with domain $D(-\Delta_{D_\pm}) = H_0^{2,2}((0, \pm\infty))$. $-\Delta_\alpha$ describes a δ-interaction of strength α centered at 0. This is the mathematical interpretation of the heuristic expression $-\frac{d^2}{dx^2} + \alpha\delta(x-y)$ (which can be also understood in the sense of sum of quadratic forms, see below).

In fact $-\Delta_\alpha$ can also be described in many alternative ways. Its resolvent is given by

$$(-\Delta_\alpha - k^2)^{-1} = G_k - 2\alpha k(i\alpha + 2k)^{-1} (\overline{G_k(\cdot)}, \cdot) \, G_k(\cdot), \quad k^2 \notin \sigma(-\Delta_\alpha), \ \text{Im}k > 0, \tag{3.1}$$

with $G_k(x,x') \equiv \frac{i}{2k} e^{ik|x-x'|}$.

As opposite to the 3-dimensional case, $-\Delta_\alpha$ can also be described as a perturbation of $-\Delta$ in the sense of quadratic forms (in the 3-dimensional case this is only possible in a generalized sense, via renormalization and limits, as we saw in Sect. 1).

Let us consider the quadratic form $Q_\alpha(g,h) \equiv (g',h') + \alpha\bar{g}(0)h(0)$, with (,) the scalar product in $L^2(\mathbb{R})$, $g,h \in D(Q_\alpha) \equiv H^{2,1}(\mathbb{R})$.

Q_α is semibounded and closed and the unique self-adjoint operator associated with it is $-\Delta_\alpha$, cfr. e.g. [30] .

Yet another possibility of describing $-\Delta_\alpha$ is by the theory of local Dirichlet forms, as in the 3-dimensional case described in Sect. 1.1 .

Let H_α be the unique self-adjoint operator associated with the closure of the local Dirichlet form in $L^2(\mathbb{R}, \varphi_\alpha^2 dx)$: $E_\alpha(g,h) \equiv \int_{\mathbb{R}} dx \varphi_\alpha^2(x)\overline{g'(x)}\, h'(x)$ on $D(E_\alpha) = C_0^1(\mathbb{R})$, with $\varphi_\alpha(x) \equiv e^{\frac{\alpha}{2}|x|}$, $\alpha \in \mathbb{R}$, $\varphi_\alpha(x) \equiv 1$ for $\alpha = +\infty$.

Then $-\Delta_\alpha = \varphi_\alpha H_\alpha \varphi_\alpha^{-1} - \frac{\alpha^2}{4}$.

<u>Remark.</u> These results extend to accretive extensions of iH corresponding to δ-interactions of strength α with Im $\alpha < 0$.

The spectrum $\sigma(-\Delta_\alpha)$ is easily determined from the formula for the resolvent:(see [11])

$\sigma(-\Delta_\alpha) = \{-\frac{\alpha^2}{2}\} \cup [0,\infty)$,

for $\alpha < 0$, $\sigma(-\Delta_\alpha) = [0,\infty)$ for $\alpha \geq 0$.

$-\frac{\alpha^2}{4}$ is a simple eigenvalue if $\alpha < 0$, with eigenfunction $(-\frac{\alpha}{2})^{1/2} e^{\alpha|x|/2}$.

There are no eigenvalues in $[0,\infty)$ nor singular continuous spectrum, and the spectrum in $[0,\infty)$ is essential and absolutely continuous.

For $\alpha \geq 0$ there is a simple resonance at $k_0 = -i\alpha/2$, with corresponding resonance function $e^{\alpha|x|/2}$, $\alpha \geq 0$.

<u>Remark.</u> Similarly as in the 3-dimensional case, it is possible to approximate $-\Delta_\alpha$ by means of scaled short range Hamiltonians with integrable potential V in the norm resolvent sense:

$\text{n-lim}_{\varepsilon \downarrow 0} (H_\varepsilon - k^2)^{-1} = (-\Delta_\alpha - k^2)^{-1}$, $k^2 \notin \sigma(-\Delta_\alpha)$,

with $H_\varepsilon = \varepsilon^{-2} U_\varepsilon H(\varepsilon) U_\varepsilon^{-1}$, $(U_\varepsilon g)(x) \equiv \varepsilon^{-1/2} g(x/\varepsilon)$, $\varepsilon > 0$, $g \in L^2(\mathbb{R})$, $H(\varepsilon) = -\Delta + \lambda(\varepsilon)\, V$, the sum being in the sense of quadratic forms. In this result we have λ real analytic near the origin, with $\lambda(0) = 0$, and $\alpha = \lambda'(0) \int_{\mathbb{R}} V(x)dx$ (see [44]).

The case $\alpha = +\infty$ cannot be reached by such limits. The convergence of eigenvalues, resonances and the scattering matrix for H_ε to the corresponding quantities for $-\Delta_\alpha$ is discussed in [11] .

B. Finitely many point interactions

Let N be a fixed natural number, and let $Y = \{y_1,\dots,y_N\}$ be a set of sources in $\mathbb{R}$. Define $\dot{H}_Y$ in $L^2(\mathbb{R})$ by $\dot{H}_Y = -\frac{d^2}{dx^2}$ with domain $D(\dot{H}_Y) \equiv \{g \in H^{2,2}(\mathbb{R}) \,|\, g(y_j) = 0,$ $j=1,\dots,N\}$.

Then $\dot{H}_Y$ is closed, symmetric and nonnegative and has deficiency indices (N,N), hence there is an N^2-parameter family of self-adjoint extensions. This family is determined by boundary conditions, we shall only consider separated boundary conditions at the points y_j, yielding local interactions. These boundary conditions give an N-parameter family of self-adjoint extensions $-\Delta_{\alpha,Y}$ of $\dot{H}_Y$:

$-\Delta_{\alpha,Y} = -\frac{d^2}{dx^2}$ with domain

$D(-\Delta_{\alpha,Y}) = \{g \in H^{2,1}(\mathbb{R}) \cap H^{2,2}(\mathbb{R}-Y) \,|\, g'(y_j+)-g'(y_j-) = \alpha_j g(y_j),\ j=1,\dots,N\}$, $\alpha = (\alpha_1,\dots,\alpha_N)$, $\alpha_j \in \mathbb{R}$.

We have $-\Delta_{o,Y} = -\Delta$ with domain $H^{2,2}(\mathbb{R})$, whereas in the case where $\alpha_{j_o} = +\infty$ for some $j_o \in \{1,\dots,N\}$ we have that $-\Delta_{\alpha,Y}$ is the Dirichlet Laplacian with Dirichlet boundary condition at y_{j_o}, and else the boundary conditions given in the definition of $D(-\Delta_{\alpha,Y})$ for $j \neq j_o$. $-\Delta_{\alpha,Y}$ describes N δ-interactions of strengths α_j centered at the points $y_j \in \mathbb{R}$, heuristically $-\Delta_{\alpha,Y} = -\frac{d^2}{dx^2} + \sum_{j=1}^{N} \alpha_j \delta(x-y_j)$ (this is also rigorous as sum of quadratic forms).

The resolvent of $-\Delta_{\alpha,Y}$ is given by

$$(-\Delta_{\alpha,Y}-k^2)^{-1} = G_k + \sum_{j,j'=1}^{N} [\Gamma_{\alpha,Y}(k)]^{-1}_{j,j'} \, (\overline{G_k(\cdot-y_{j'})},\cdot)\, G_k(\cdot-y_j), \quad k^2 \notin \sigma(-\Delta_{\alpha,Y}), \tag{3.2}$$

$\mathrm{Im}\, k > 0$, $-\infty < \alpha_j \leq +\infty$, where $[\Gamma_{\alpha,Y}(k)]_{jj'} = -\alpha_j \delta_{jj'} - G_k(y_j-y_{j'})$.

It is understood that if $\alpha_{j_o} = 0$ one deletes the j_o-th line and j_o-th row in the matrix $\Gamma_{\alpha,Y}$.

<u>Remark.</u> If for at most one j_o one has $\alpha_{j_o} = +\infty$ and $\alpha_j \neq 0 \ \forall j \in \{1,\dots,N\}$, then $-\Delta_{\alpha,Y}$ has at most N eigenvalues which are all negative and simple. If $\alpha_j = +\infty$ for at least two different values of j then $-\Delta_{\alpha,Y}$ has at most N eigenvalues (counting multiplicity) and infinitely many eigenvalues embedded in $[0,\infty)$,

accumulating at $+\infty$. The remaining part of the spectrum is absolutely continuous and covers $[0,\infty)$. For these results see [11], [44].

As in the case of one center $-\Delta_{\alpha,Y}$ can be described as a perturbation of $-\Delta$ in the sense of quadratic forms: $-\Delta_{\alpha,Y}$ is the unique self-adjoint operator associated with the semibounded, closed quadratic form

$$Q_{\alpha,Y}(g,h) \equiv (g',h') + \sum_{j=1}^{N} \alpha_j \overline{g(y_j)}\, h(y_j), \text{ with domain } D(Q_{\alpha,Y}) = H^{2,1}(\mathbb{R}).$$

<u>Remark.</u> As for one center one gets norm resolvent convergence of $H_{\varepsilon,Y}$ to $-\Delta_{\alpha,Y}$, with $H_{\varepsilon,Y} \equiv -\Delta + \sum_{j=1}^{N} \lambda_j(\varepsilon)\, V_j(\cdot - \frac{1}{\varepsilon} y_j)$, $\varepsilon > 0$, $V_j \in L^1(\mathbb{R})$, λ_j real analytic near 0 with $\lambda_j(0) = 0$, $\alpha_j = \lambda_j'(0) = \int_{\mathbb{R}} dx\, V_j(x)$ (see [44]).

This then yields results on the corresponding convergence of eigenvalues, resonances and the scattering matrix.

<u>C. Infinitely many point interactions.</u>

Let $J \subset \mathbb{Z}$ be an infinite index set. Let $Y = \{y_j \in \mathbb{R} \mid j \in J\}$ be a discrete subset of $\mathbb{R}$ s.t. for some $d > 0$

$$\inf_{j,j' \in J, j \neq j'} |y_j - y_{j'}| \geq d > 0, \quad y_j, y_{j'} \in Y.$$

Assume (for notational convenience) $j \in J \to j+1 \in J$ and $y_j < y_{j+1}$.

Let $\dot{H}_y = -\frac{d^2}{dx^2}$ with domain $D(\dot{H}_y) = \{g \in H^{2,2}(\mathbb{R}) \mid g(y_j) = 0 \ \forall j \in J\}$.

Then $\dot{H}_y$ is closed, nonnegative and has deficiency indices (∞,∞) (cfr. [11]).

A particular class of self-adjoint extensions of $\dot{H}_y$ is the one parameter family $-\Delta_{\alpha,Y}$ given by $-\Delta_{\alpha,Y} = -\frac{d^2}{dx^2}$ with domain $D(-\Delta_{\alpha,Y}) = (g \in H^{2,1}(\mathbb{R}) \cap H^{2,2}(\mathbb{R} - Y) \mid g'(y_j+) - g'(y_j-) = \alpha_j\, g(y_j),\ j \in J\}$.

with $\alpha = \{\alpha_j,\ j \in J\}$, $-\infty < \alpha_j \leq +\infty$. $-\Delta_{\alpha,Y}$ describes point interactions of strength α_j and sources $y_j \in Y$, $j \in J$, heuristically $-\Delta_{\alpha,Y} = -\frac{d^2}{dx^2} + \sum_{j \in J} \alpha_j\, \delta(x - x_j)$ (this holds actually in the sense of limits of sums of quadratic forms, see below). For $\alpha_j = 0\ \forall j \in J$ we have $-\Delta_{\alpha,Y} = -\Delta$ with domain $H^{2,2}(\mathbb{R})$. For $\alpha_{j_o} = +\infty$ for some $j_o \in J$ we have that in the definition of $D(-\Delta_{\alpha,Y})$ the boundary condition at y_{j_o} should be interpreted as Dirichlet boundary condition i.e. $g(y_{j_o}+) = g(y_{j_o}-) = 0$.

The following approximation results, similar to the one established in 3 dimensions, is useful to get control on eigenvalues, resonances and scattering quantities for $-\Delta_{\alpha,Y}$:

<u>Theorem 11</u>. For $\alpha_j \in \mathbb{R}$, $M,N \in \mathbb{N}$ define $Y_{M,N} \subset Y$, $\alpha_{M,N} < \alpha$ with

$\alpha_{M,N} = \{\alpha_j, \{-M \le j \le N\}\}$, $Y_{M,N} \equiv \{y_j \in Y \mid -M \le j \le N\}$.

Then $(-\Delta_{\alpha_{M,N},Y_{M,N}} - k^2)^{-1} \underset{M,N\to\infty}{\longrightarrow} (-\Delta_{\alpha,Y} - k^2)^{-1}$, with strong convergence for

$k^2 \notin \sigma(-\Delta_{\alpha,Y}) \cap \rho_b$, $\mathrm{Im}\, k > 0$,

$\rho_b \equiv \{z \in \mathbb{C} \mid \exists N_o(z) \in \mathbb{N}, \exists\, C \ge 0,\ z \notin \sigma(-\Delta_{\alpha_{M,N},Y_{M,N}})$ and

$\|(-\Delta_{\alpha_{M,N},Y_{M,N}} - z)^{-1}\| \le C\ \forall M,N \ge N_o(z)\}$. $\sigma(-\Delta_{\alpha_{M,N},Y_{M,N}})$ is asymptotically concentrated on any open Borel set G s.t. $G \supset \sigma(-\Delta_{\alpha,Y})$ and, calling $P_{-\Delta_{\alpha_{M,N},Y_{M,N}}}$ resp. $P_{-\Delta_{\alpha,Y}}$ the spectral projections associated with $-\Delta_{\alpha_{M,N},Y_{M,N}}$ resp. $-\Delta_{\alpha,Y}$, we have that $P_{-\Delta_{\alpha_{M,N},Y_{M,N}}}(a,b)$ converges strongly as $M,N \to \infty$ to $P_{-\Delta_{\alpha,Y}}(a,b)$, for any a,b not in the point spectrum of $-\Delta_{\alpha,Y}$.

<u>Remark.</u> For the proof see [11] . The resolvent cannot converge in norm, in general, since $\sigma_{ess}(-\Delta_{\alpha_{M,N'},Y_{M,N}}) = [0,\infty)$, whereas one expects gaps in the positive spectrum of $-\Delta_{\alpha,Y}$. The resolvent of $-\Delta_{\alpha,Y}$ is given, for $\alpha_j \in \mathbb{R}-\{0\}$, $j \in J$, similarly as in the 3-dimensional case, by

$$(-\Delta_{\alpha,Y}-k^2)^{-1} = G_k + \sum_{j,j' \in J} [\Gamma_{\alpha,Y}(k)]^{-1}_{j,j'}\, \overline{G_k(\cdot - y_{j'},\cdot)} G_k(\cdot)\quad k^2 \notin \sigma(-\Delta_{\alpha,Y}),\ \mathrm{Im}\, k > 0,$$

where $[\Gamma_{\alpha,Y}(k)]_{j,j'} \equiv [-\alpha_j^{-1}\delta_{j,j'} - G_k(y_j - y_{j'})]_{j,j'}$ (3.3)

For the proof we refer to [44] , [11].

The following approximation result by scaled Hamiltonians with integrable potential holds and extends the corresponding result for finitely many centers. Let V_i be real valued in $L^1(\mathbb{R})$ and such that $\exists\, W \in L^1(\mathbb{R})$, $|V_j| \le W$. Let λ_j be s.t. $|\varepsilon^{-1}\lambda_j(\varepsilon)| \le C\ \forall\, 0 < \varepsilon < \varepsilon_o$, $\lambda_j(\varepsilon) = \varepsilon\lambda_j'(0) + O(\varepsilon)$ as $\varepsilon \downarrow 0$.

Then $H_{\varepsilon,Y}$ converges to $-\Delta_{\alpha,Y}$ in norm resolvent sense, with $\alpha_j \equiv \lambda_j'(0) \int_{\mathbb{R}} dx V_j(x)$.

For this result see [44], [11].

D. Periodic point interactions (Kronig-Penney model)

Let us consider the so called Kronig-Penney model. Let $\Lambda = \{n\delta \mid n \in \mathbb{Z}\}$, $\delta > 0$,

let $\hat{\Gamma} \equiv [-\delta/2, \delta/2]$ ("Wigner-Seitz cell"),

$\Gamma \equiv \{n2\pi/\delta, n \in \mathbb{Z}\}$ ("dual lattice"),

$\hat{\Lambda} \equiv [-\pi/\delta, \pi/\delta]$ ("Brilliouin zone").

The Hilbert space $L^2(\mathbb{R})$ can be decomposed as $L^2(\mathbb{R}) = U^{-1}L^2(\hat{\Lambda}; \ell^2(\Gamma)) \equiv U^{-1}\int_{\hat{\Lambda}}^{\oplus} d\theta \ell^2(\Gamma)$, where U is the mapping from $L^2(\mathbb{R})$ into $L^2(\hat{\Lambda}; \ell^2(\Gamma))$ given by $(Uf)(\theta,n) \equiv \hat{f}(\theta+nb)$, $\theta \in \hat{\Lambda}$, $n \in \mathbb{Z}$.

The Fourier transform picture is

$$L^2(\mathbb{R}) = \tilde{U}^{-1}L^2(\hat{\Lambda}, \delta d\theta/2\pi; L^2(\hat{\Gamma})) \equiv \tilde{U}^{-1}\int_{\hat{\Lambda}}^{\oplus} \delta d\theta/2\pi\, L^2(\hat{\Gamma}),$$

where $\tilde{U}$ is the map into $L^2(\hat{\Lambda}, \delta d\theta/2\pi; L^2(\hat{\Gamma}))$, given for $f \in \mathcal{S}(\mathbb{R})$ by

$$(\tilde{U}f)(\theta,\nu) \equiv \sum_{n\in\mathbb{Z}} e^{-in\theta a} f(\nu + na).$$

θ is the so called "quasi momentum" or "Bloch's vector".

Let $\alpha \equiv \alpha_j$ $\forall j \in \mathbb{Z}$, $Y = \Lambda$ and write $-\Delta_{\alpha,\Lambda}$ for $-\Delta_{\alpha,Y}$ in this case.

Define the operators in $L^2(-\frac{a}{2}, \frac{a}{2})$, $-\Delta_{\alpha,\Lambda}(\theta) \equiv -\frac{d^2}{d\nu^2}$ with domain

$$D(-\Delta_{\alpha,\Lambda}(\theta)) = \{g(\theta) \in H^{2,1}((-\tfrac{\delta}{2}, \tfrac{\delta}{2})) \cap H^{2,2}((-\tfrac{\delta}{2}, \tfrac{\delta}{2}) - \{0\}) \mid$$

$$g(\theta, \tfrac{\delta}{2}+) = e^{i\theta\delta} g(\theta, \tfrac{\delta}{2}-),\ g'(\theta, \tfrac{\delta}{2}+) = e^{i\theta\delta} g'(\theta, \tfrac{\delta}{2}+),$$

$$g(\theta,0+) = g(\theta,0-),\ g'(\theta,0+) - g'(\theta,0+) = \alpha g(\theta,0)\},\ -\infty < \alpha \leq +\infty.$$

The following result determines the structure of $-\Delta_{\alpha,\Lambda}(\theta)$.

Theorem 12 (cfr. [11]):

For any $-\infty < \alpha \leq +\infty$, $\theta \in \hat{\Lambda}$,

the spectrum of $-\Delta_{\alpha,\Lambda}(\theta)$ is purely discrete, consisting of eigenvalues $E_m^{\alpha,\Lambda}(\theta)$, $m \in \mathbb{N}$ given by $E_m^{\alpha,\Lambda}(\theta) = [k_m^{\alpha,\Lambda}(\theta)]^2$, $m \in \mathbb{N}$, where $k_m^{\alpha,\Lambda}(\theta)$ are solutions of the Kronig-Penney relations

$\cos\theta\delta = \cos k\delta + \frac{\alpha}{2k}\sin k\delta$, $\operatorname{Im} k \geq 0$. $E_m^{\alpha,\Lambda}(\theta)$, $\alpha \in \mathbb{R} - \{0\}$, $\theta \in \hat{\Lambda}$ are non degenerate for $\alpha \neq 0$ and

$$0 < E_1^{\alpha,\Lambda}(0) < E_1^{\alpha,\Lambda}(-\pi/\delta) = \pi^2/\delta^2 < E_2^{\alpha,\Lambda}(-\pi/\delta) < E_2^{\alpha,\Lambda}(0) =$$

$$= 4\pi^2/\delta^2 < E_3^{\alpha,\Lambda}(0) < E_3^{\alpha,\Lambda}(-\pi/\delta) = 9\pi^2/\delta^2 < E_5^{\alpha,\Lambda}(0) < \ldots \ \forall \alpha > 0;$$

$$E_1^{\alpha,\Lambda}(0) < E_1^{\alpha,\Lambda}(-\pi/\delta) < E_2^{\alpha,\Lambda}(-\pi/\delta) = \pi^2/\delta^2 < E_2^{\alpha,\Lambda}(0) < E_3^{\alpha,\Lambda}(0) = 4\pi^2/\delta^2 < \ldots,$$

$$E_1^{\alpha,\Lambda}(0) < 0,\ E_1^{\alpha,\Lambda}(-\pi/\delta) = \begin{cases} < 0 & \text{if } -\alpha > 4/\delta \\ = 0 & \text{if } -\alpha = 4/\delta \\ < 0 & \text{if } -\alpha < 4/\delta \end{cases},$$

for $\alpha < 0$.

For $\alpha = 0$ the eigenvalues are simply

$$E_{m,\pm}^{0,\Lambda}(\theta) = \{\pm\theta + [2(m-1)\pi/\delta]\}^2,\ m \in \mathbb{N}.$$

They are only degenerate for $\theta = -\pi/\delta$ and $\theta = 0$. For $\alpha = +\infty$ we have $E_m^{0,\Lambda}(\theta) = m^2\pi^2/\delta^2\ \forall\theta$, and these eigenvalues are simple.

Proof. See [11].

This result implies, by the fact that $\tilde{U}[-\Delta_{\alpha,\Lambda}]\tilde{U}^{-1} = \int_{\hat{\Lambda}}^{\oplus} \delta d\theta/2\pi\, [-\Delta_{\alpha,\Lambda}(\theta)]$ the following result on the spectrum of $-\Delta_{\alpha,\Lambda}$:

Theorem 13. For any $\alpha \in \mathbb{R}$, $-\Delta_{\alpha,\Lambda}$ has purely absolutely continuous spectrum

$$\sigma(-\Delta_{\alpha,\Lambda}) = \bigcup_{m=1}^{\infty} [a_m^{\alpha,\Lambda}, b_m^{\alpha,\Lambda}],\ \text{with } a_m^{\alpha,\Lambda} < b_m^{\alpha,\Lambda} \leq a_{m+1}^{\alpha,\Lambda},\ m \in \mathbb{N}.$$

For $\alpha > 0$: $a_1^{\alpha,\Lambda} > 0$, $a_m^{\alpha,\Lambda} = \begin{cases} E_m^{\alpha,\Lambda}(0), & m \text{ odd} \\ E_m^{\alpha,\Lambda}(-\pi/\delta), & m \text{ even} \end{cases}$,

$$a_m^{\alpha,\Lambda} > (m-1)^2\pi^2/\delta^2;\ b_m^{\alpha,\Lambda} = \begin{cases} E_m^{\alpha,\Lambda}(-\pi/\delta) = m^2\pi^2/\delta^2, & m \text{ odd} \\ E_m^{\alpha,\Lambda}(0) = m^2\pi^2/\delta^2, & m \text{ even} \end{cases}.$$

For $\alpha < 0$: $a_1^{\alpha,\Lambda} = E_1^{\alpha,\Lambda}(0) < 0$; $b_1^{\alpha,\Lambda} = E_1^{\alpha,\Lambda}(-b/2) = \begin{cases} < 0 \text{ if } |\alpha| > 4/\delta \\ = 0 \text{ if } |\alpha| = 4/\delta \\ > 0 \text{ if } |\alpha| < 4/\delta \end{cases}$

$$a_m^{\alpha,\Lambda} \equiv \begin{cases} E_m^{\alpha,\Lambda}(0) = (m-1)^2\pi^2/\delta^2, & m \text{ odd} \geq 2 \\ E_m^{\alpha,\Lambda}(-b/2) = (m-1)^2\pi^2/\delta^2, & m \text{ even} \end{cases},$$

$$b_m^{\alpha,\Lambda} \equiv \begin{cases} E_m^{\alpha,\Lambda}(-\pi/\delta), & m \text{ odd},\ m \geq 2;\ b_m^{\alpha,\Lambda} < m^2\pi^2/\delta^2,\ m \geq 2 \\ E_m^{\alpha,\Lambda}(0), & m \text{ even} \end{cases}$$

with $E_m^{\alpha,\Lambda}(\theta)$ the eigenvalues of $-\Delta_{\alpha,\Lambda}(\theta)$ described in Theorem 12. a_n^{α}, b_m^{α} are continuous with respect to $\alpha \in \mathbb{R}$.

For $\alpha \in \mathbb{R} - \{0\}$ all possible (infinitely many) gaps in $\sigma(-\Delta_{\alpha,\Lambda})$ occurr. For $\alpha = 0$ the spectrum is the one, $[0,\infty)$, of $-\Delta$ on $H^{2,2}(\mathbb{R})$ (all gaps closed). For $\alpha = +\infty$, $-\Delta_{\alpha,\Lambda}$ is the Dirichlet Laplacian on $\mathbb{R} - \Lambda$, i.e. $\sigma(-\Delta_{\alpha,\Lambda})$ is pure point with eigenvalues $m^2\pi^2/\delta^2$, $m \in \mathbb{N}$ of infinite multiplicity. $\sigma(-\Delta_{\alpha,\Lambda})$ has the following monotonicity properties: $\sigma(-\Delta_{\alpha,\Lambda}) \subset \sigma(-\Delta_{\alpha',\Lambda})$ for $0 \leq \alpha' < \alpha$; $\sigma(-\Delta_{\alpha,\Lambda}) \cap [0,\infty) \supset$ $\supset \sigma(-\Delta_{\alpha',\Lambda}) \cap [0,\infty)$, $\alpha' < \alpha \leq 0$.

Proof. See [43], [11].

Remark. In [11] extensions of this result to the case where (α,Λ) is such that $\alpha \equiv \alpha^{(p)} \equiv (\alpha_j^{(p)}, j \in \mathbb{Z})$, $\alpha_{j+p}^{(p)} = \alpha_j^{(p)}$ $\forall_j \in \mathbb{Z}$ are given.

Such extensions use in particular a connection between $-\Delta_{\alpha,Y}$ in $L^2(\mathbb{R})$ and a certain discrete operator in $\ell^2(\mathbb{Z})$, studied in [45].

The above results have applications to the study of 1-dimensional alloys. Let us consider an alloy consisting of N different sorts of "atoms" represented by point interactions with equally spaced sources with distance $a > 0$ between each other of strengths $\gamma_n \in \mathbb{R}$, $\gamma_n \neq \gamma_{n'}$ for $n \neq n', n, n' = 1,\ldots,N$, arranged in the following way: the primitive cell $\hat{\Gamma}$ of the model consists of p_1 points supporting point interations of strength γ_{n_1}, followed by p_2 points supporting point interactions of strength γ_{n_2},...,up to p_M points supporting point interactions of strength γ_{n_M}, with $M \geq N$. Denote the corresponding Hamiltonian by $-\Delta_{\alpha_p, Y_p+\Lambda_p}$, where $\alpha_p = (\ldots\{\gamma_{n_i},\ldots\gamma_{n_i}\}\ldots)$, γ_{n_i} occurring p_i-times, $i = 1,\ldots,M$, $\sum_{i=1}^{M} p_i = p$, $Y_p = \{n\delta \mid 0 \leq n \leq p-1\}$, $\Lambda_p = p\,\delta\,\mathbb{Z}$.

Then we have the

Theorem 14. Let ρ_N be the intersection of all spectral gaps of the Hamiltonians $-\Delta_{\gamma_n,\delta\mathbb{Z}}$ of the pure Kronig-Penney crystals with potential strength $\gamma_n \in \mathbb{R}$ and fixed Bravais lattice $\delta\mathbb{Z}$, $\delta > 0$ i.e. $\rho_N = \bigcap_{n=1}^{N} \rho(-\Delta_{\gamma_n,\delta\mathbb{Z}})$, with $\rho(A)$ the resolvent set of the operator A. Then $\dot{\rho}_N \subset \rho(-\Delta_{\alpha_p, Y_p+\Lambda_p})$.

Remark. This means that common spectral gaps of all pure crystals described by $-\Delta_{\gamma_n,\delta\mathbb{Z}}$ remain spectral gaps for all alloys represented by $-\Delta_{\alpha_p, Y_p+\Lambda_p}$, $p \in \mathbb{N}$ consisting of N sorts of atoms.

Hence this proves a much more general form of the original so called Saxon-Hutner conjecture (a naive generalization of this conjecture to non-point interactions does not hold, for a correct generalization see [6], [20]).

Proof. Cfr. [47] (previous results had been obtained in [20], [7], [21], [49]).

E. Random points interactions in one dimension

Let us consider as before the lattice $\Lambda = \mathbb{Z}$, with point interactions of random strength $\alpha_i(\omega)$, $i \in \Lambda$. We assume for simplicity (see remark at the end for extensions) that α_j are identically distributed independent random variables on the canonical probability space $(\Omega, \mathcal{A}, P)$ and $|\alpha_j(\omega)| \leq C \; \forall j \in \mathbb{Z}$, $\omega \in \Omega$. Thus $\Omega = [-C,+C]^{\mathbb{Z}}$ is a compact topological space, hence we many speak of the support of the probability measure P. Let us consider, in the notation of D, $H_\omega = -\Delta_{\alpha(\omega),\Lambda}$.

Let, as in the 3-dimensional case, Sect. 2 , T_j be the mapping of Ω into itself given by $\alpha_j(T\omega) = \alpha_{j-1}(\omega)$.

As in Sect. 2 , T is ergodic and $H_{T_\omega} = U H_\omega H^{-1}$, with $(Uf)(x) \equiv f(x-a)$, $\forall f \in L^2(\mathbb{R})$. correspondingly as in Sect. 2 one shows

Theorem 15. The spectrum $\sigma(H_\omega)$ of H_ω and its parts $\sigma_{ac}(H_\omega)$, $\sigma_{sc}(H_\omega)$, $\sigma_{pp}(H_\omega)$ are all P-a.s. equal to non random sets Σ resp. Σ_{ac}, Σ_{sc}, Σ_{pp}. The discrete spectrum of H_ω is void, P-a.s. Any $\lambda \in \mathbb{R}$ is with probability zero an eigenvalue of H_ω.

Proof. For the proof we have to verify the measurablity of H_ω in the sense of Sect. 2 and this follows easily from the formula (3.3) for the resolvent of H_ω, together with the approximation theorem 11 and the measurability of $\Gamma_{\alpha(\omega),Y}(k)$ (with $Y = \Lambda$) in (3.3). For details see [43], [8].

Let us now investigate in more details the structure of $\sigma(H_\omega)$. Similarly as for the 3-dimensional case the approximation of H_ω by Hamiltonians with periodic potentials will be the main tool.

Let us call a sequence $\Phi(\omega) = \{\alpha_i(\omega),\; i \in \mathbb{Z}\}$ a stochastic potential (we do not need to introduce the stochastic variables $X_i(\omega)$ here, since $\alpha_i(\omega) = 0$ is possible, if $P(\alpha_i = 0) > 0$).

An admissible potential is $\varphi = (\lambda_i, i \in \mathbb{Z})$, with $\lambda_i \in \text{supp } P_{\alpha_o}$ $\forall i \in \mathbb{Z}$. Let $\mathfrak{A}$ be the class of all admissible potentials and let $\mathfrak{P}$ be the class of all periodic admissible potentials, i.e. $\varphi \in \mathfrak{P}$ if $\varphi \in \mathfrak{A}$ and $\lambda_{i+n} = \lambda_i$ for some $n \in \mathbb{N}$. For $\varphi \in \mathfrak{A}$ we denote by $H(\varphi)$ the Hamiltonian H_ω with $\alpha_i(\omega)$ replaced by λ_i.

<u>Theorem 16.</u> For any $\omega \in \text{supp } P$ we have $\sigma(H_\omega) \subset \Sigma$, Σ as in Theorem 15.
In particular, for any $\varphi \in \mathfrak{A}$,

i) $\sigma(H(\varphi)) \subset \Sigma$. Moreover:

ii) $$\Sigma = \bigcup_{\varphi \in \mathfrak{A}} \sigma(H(\varphi)) = \overline{\bigcup_{\varphi \in \mathfrak{P}} \sigma(H(\varphi))}.$$

<u>Proof.</u> We know from Theorem 15 that $\Omega^1 \equiv \{\omega \mid \sigma(H_\omega) = \Sigma\}$ has probability 1. If $\varphi \in \mathfrak{A}$ we contruct a sequence $\omega_n \in \Omega^1$ s.t. H_{ω_n} converges in the strong resolvent sense towards H_φ. This is done, similarly as in the proof of Theorem 6, by introducing Ω_n^φ with $P(\Omega_n^\varphi) > 0$, hence $\Omega^1 \cap \Omega_n^\varphi \neq \emptyset$.

Let $Y(\varphi) = \{i \in \mathbb{Z} \mid \lambda_i > 0\}$ and denote by $T_k(\varphi)$ the operator $\Gamma_{\alpha,Y}$ of (3.3) restricted to $\ell^2(Y(\varphi))$ and set

$$\tilde{T}_k(\varphi) \equiv T_k(\varphi) + \mathbb{1}_{Y(\varphi)^c}, \quad Y(\varphi)^c \equiv \mathbb{Z} - Y(\varphi).$$

We then have, with $T_k(\omega_n) \equiv T_k(\varphi_n)$ with $\varphi_n \equiv (\alpha_i(\omega_n), i \in \mathbb{Z})$,
$\tilde{T}_k(\omega_n) \equiv \tilde{T}_k(\varphi_n)$, $Y(\omega_n) \equiv Y(\varphi_n)$, that $\tilde{T}_k(\varphi_n)^{-1} - \mathbb{1}_{Y(\omega_n)^c} \to T_k(\varphi)^{-1} - \mathbb{1}_{Y(\varphi)^c}$ strongly in $\ell^2(\mathbb{Z})$, which implies the strong convergence of H_{ω_n} to $H(\varphi)$. This then yields i).

ii) is proven entirely similarly as in the proof of Theorem 6. □

<u>Remark.</u> The above result can be obtained by exploiting the definition of one dimensional point interactions via quadratic forms [6,20]. This proof, however, has no higher dimensional counterpart.

Let us set $H_\lambda \equiv -\Delta_{\lambda,\mathbb{Z}}$. Then we have:

<u>Theorem 17.</u> Let Σ be as in Theorem 15. Let supp P_{α_o} be compact, then

$$\Sigma = \bigcup_{\lambda \in \text{supp } P_o} \sigma(H_\lambda) \text{ with } P_{\alpha_o} \equiv P_o \text{ the distribution of } \alpha_o.$$

Proof. We have $H_\lambda \equiv -\Delta_{\lambda,\mathbb{Z}} = H(\varphi)$ with $\varphi = \{\lambda_i = \lambda \ \forall i \in \mathbb{Z}\}$. By Theor. 16 we then have

$$\Sigma_o \equiv \bigcup_{\lambda \in \text{supp } P_o} \sigma(H_\lambda) \subset \Sigma .$$

Conversely, there is a $\varphi = (\lambda_i,\ i \in \mathbb{Z}) \in \mathfrak{A}$ such that $\Sigma = \sigma(H(\varphi))$ (this is even true for P-a.e. $\omega = (\lambda_i,\ i \in \mathbb{Z}) \in \Omega$). By the theorem 14 if $(c,d) \cap \Sigma_o = \emptyset$ then also $(c,d) \cap \bar{\Sigma} = \emptyset$. Thus $\Sigma \subset \bar{\Sigma}_o$.

We claim that Σ_o is closed.

In fact by Theorem 13 , for any $\alpha \in \mathbb{R}$

$$\sigma(H_\alpha) = \sigma_{ac}(H_\alpha) = \bigcup_{m=1}^{\infty} [a_m^\alpha, b_m^\alpha],\ a_m^\alpha < b_m^\alpha \le a_{m+1}^\alpha,\ m \in \mathbb{N},\ \sigma_{sc}(H_\alpha) = \sigma_{pp}(H_\alpha) = \emptyset .$$

Thus $\mathcal{A}_m \equiv \bigcup_{\lambda \in \text{supp } P_o} [a_m^\lambda, b_m^\lambda] \subset [a_m^{\lambda_-}, b_m^{\lambda_+}]$, for some $\lambda_\pm \in \text{supp } P_o$, by the assumed compactness of supp P_o and the continuity of a_m^λ, b_m^λ in $\lambda \in \mathbb{R}$.

Suppose the sequence $\{x_n\}$, $n \in \mathbb{N}$, $\{x_n\} \in \mathcal{A}_m$ converges as $n \to \infty$. Then $\exists x_o \in [a_m^{\lambda_-}, b_m^{\lambda_+}]$ s.t. $x_o = \lim x_n$, and $x_n \in [a_m^{\lambda_n}, b_m^{\lambda_n}]$ for some λ_n. By the compactness of supp P_o, λ_n has a subsequence $\lambda_{n'}$ s.t. $\lambda_o \equiv \lim_{n' \to \infty} \lambda_{n'} \in \text{supp } P_o$.

From $a_m^{\lambda_n} \le x_n \le b_m^{\lambda_n}$ we get by continuity $a_m^{\lambda_o} \le x_o \le b_m^{\lambda_o}$, hence $x_o \in \mathcal{A}_m$. This proves $\mathcal{A}_n$ is closed. From the restrictions on the a_m, b_m given by Theorem 13 we then get $\bigcup_{m=1}^{\infty} \mathcal{A}_m = \Sigma_o$, hence Σ_o is closed. Thus $\Sigma_o = \Sigma$, which proves the theorem. □

Corollary: $\Sigma = \bigcup_{m=1}^{\infty} [a_m, b_m]$ where $a_m < b_m \le a_{m+1}$ and $a_m, b_m \to \infty$ as $m \to \infty$.

Proof. From the above we have $\Sigma = \bigcup_{\lambda \in \text{supp } P_o} \bigcup_{m=1}^{\infty} [a_m^\lambda, b_m^\lambda]$, with a_m^λ, b_m^λ satisfying

From this the Corollary follows.

Corollary: If supp $P_o \subset [0,\infty)$ then $\Sigma = \sigma(H_{\tilde{a}})$, with $\tilde{a} \equiv \inf \text{supp } P_o$.

$\sigma(H_{\tilde{a}}) = \sigma(-\Delta_{\tilde{a};\mathbb{Z}})$ is given in Theorem 13 .

Proof. This follows from the above results on a_m^λ, b_m^λ for $\lambda > 0$.

Corollary: Let inf supp $P_o = \tilde{a}$, sup supp $P_o = \tilde{b}$.

i) If $\tilde{a} > 0$ then $\Sigma = \sigma(H_{\tilde{a}}) = \bigcup_{n=1}^{\infty} [a_n^{\tilde{a}}, b_n^{\tilde{a}}]$;

ii) If $\tilde{b} < 0$ and if for any $E \in [a_1(\tilde{a}), b_1(\tilde{b})]$ $\exists$ $\lambda \in \text{supp } P_o$ with $a_1(\lambda) \leq E \leq b_1(\lambda)$ (which is e.g. the case when $\text{supp } P_o = [a,b]$ or if $b_1(\tilde{a}) \geq a_1(\tilde{b})$), then

$$\Sigma = [a_1(\tilde{a}), b_1(\tilde{b})] \cup \bigcup_{n=2}^{\infty} [a_n(\tilde{b}), b_n(\tilde{b})].$$

Proof:

a) $\tilde{a} > 0$. In this case from Theorem 13

$$\sigma(H_{\tilde{a}}) = \bigcup_{n=1}^{\infty} [a_n(\lambda), b_n(\lambda)] = \bigcup_{n=1}^{\infty} [a_n(\lambda), (\frac{2n\pi}{p})^2].$$

But $a_n(\lambda)$ is monotonically decreasing in λ, by the minimax principle, and $b_n(\lambda) = [2n\pi/p]^2$ is independent of λ, so that

$$\sigma(H_\lambda) \subset \sigma(H_{\tilde{a}})$$

for any $\lambda \geq \tilde{a}$.

By Theorem 17 $\Sigma \subset \sigma(H_{\tilde{a}})$.

On the other hand by Theorem 16

$$\sigma(H_{\tilde{a}}) \subset \Sigma \text{ , hence } \Sigma = \sigma(H_{\tilde{a}}).$$

b) Let $\tilde{b} < 0$. In this case by Theorem 13

$$\sigma(H_{\tilde{a}}) = [a_1(\lambda), b_1(\lambda)] \cup \bigcup_{n=2}^{\infty} [(\pi(n-1)/p)^2, b_n(\lambda)],$$

with $a_1(\tilde{a}) \leq a_1(\lambda) < 0$, $b_n(\lambda) \leq b_n(\tilde{b})$, for any $\tilde{a} \leq \lambda \leq \tilde{b}$, $n \in \mathbb{N}$, $b_n(\lambda)$ strictly monotone increasing, continuous in λ, $b_1(\lambda_o) = 0$ for some $\lambda_o < 0$.

Thus, for any $\lambda \in [\tilde{a}, \tilde{b}]$

$$\sigma(H_\lambda) = [a_1(\tilde{a}), b_1(\tilde{b})] \cup \bigcup_{n=2}^{\infty} [(2\pi(n-1)/p)^2, b_n(\tilde{b})].$$

Using previous results $\Sigma = \bigcup_{\lambda \in \text{supp } P_o} \sigma(H_\lambda)$, with $\sigma(H_\lambda)$ given by the above.

But $\sigma(H_{\tilde{b}})^+ = [0, b_1(\tilde{b})] \cup \bigcup_{n=2}^{\infty} [[2\pi(n-1)/p]^2, b_n(\tilde{b})]$, where + means positive part.

By the monotonicity then $\sigma(H_\lambda)^+ = \sigma(H_{\tilde{b}})^+$. □

Remark. In the situation of this corollary the spectrum of H_a contains infinitely many gaps. This theorem extends in particular the Saxon-Hutner conjecture and the above Theorem to the random case. We have

Theorem 18. Let $\{\alpha_i^\omega, i \in \mathbb{Z}\}$ be an i.i.d. family, with $\tilde{a} \equiv \inf \text{supp } P_{\alpha_o}$, $\tilde{b} \equiv \sup \text{supp } P_{\alpha_o}$

Assume $0 \leq \tilde{a} \leq \tilde{b} < \infty$ or $-\infty < \tilde{a} \leq \tilde{b} \leq 0$.

Let $\{x_i, i \in \mathbb{Z}\}$ be a lattice in $\mathbb{R}$. Denote by $\Gamma_n(\lambda)$ the n-th gap of the operator $H_\lambda = -\Delta_{\lambda,\mathbb{Z}}$, i.e. $\Gamma_n(\lambda) \equiv \{x \in \mathbb{R} | y < x < z \ \ \forall y \in B_n, \forall z \in B_{n+1}\}$ with $B_n \equiv \{E_n(\theta) | \theta \in [0,\pi]\}$, $E_n(\theta)$ the n-th eigenvalue, counting multiplicity, of the reduced Hamiltonian $H_\lambda(\theta)$. Then for any open set Γ s.t. $\Gamma \subset \Gamma_n(a) \cap \Gamma_n(b)$ for some $n \in \mathbb{N}$ we have $\Gamma \cap \Sigma = \emptyset$.

Proof:a) $\tilde{a} > 0$. In this case from the Corollary we have $\Sigma = \bigcup_{n=1}^{\infty} [a_n(\tilde{a}), b_n(\tilde{a})]$.

But $b_n(\lambda)$ being independent of λ we have in particular $b_n(\tilde{a}) = b_n(\tilde{b})$.

Thus $\Sigma = \bigcup_{n=1}^{\infty} [a_n(\tilde{a}), b_n(\tilde{b})]$. Thus $\Sigma = \bigcup_{n=1}^{\infty} [a_n(\tilde{a}), b_n(\tilde{b})]$. This shows that

$\Gamma \subset \Gamma_n(\tilde{a}) \cap \Gamma_n(\tilde{b}) \Rightarrow \Gamma \cap \Sigma = \emptyset$.

b) $\tilde{b} < 0$. In this case from the theorem

$$\Sigma \subset [a_1(\tilde{a}), b_1(\tilde{b})] \cup \bigcup_{n=2}^{\infty} [(2\pi(n-1)/p)^2, b_n(\tilde{b})].$$

On the other hand

$\Gamma_n(\tilde{a}) \subset (-\infty, a_1(\tilde{a})) \cup [b_1(\tilde{a}), (2\pi(n-1)/p)^2] \cup [b_n(\tilde{a}), \infty)$,

$\Gamma_n(\tilde{b}) \subset (-\infty, a_1(\tilde{b})] \cup [b_1(\tilde{b}), (2\pi(n-1)/p)^2] \cup [b_n(\tilde{b}), \infty)$.

Thus also in this case, using $a_1(\tilde{a}) \leq a_1(\tilde{b})$, $b_n(\tilde{b}) \geq \lim b_n(\tilde{a})$,

$\Gamma \subset \Gamma_n(\tilde{a}) \cap \Gamma_n(\tilde{b})$, thus $\Gamma \cap \Sigma = \emptyset$. ◻

The Saxon-Hutner conjecture is true in the following more general form proven by Englisch [21] and in [47] :

Theorem: Suppose that the open set Γ is contained in $\mathbb{R} \setminus \sigma(H_\lambda)$ for all $\lambda \in \operatorname{supp} P_o$. Then $\Gamma \cap \sigma(H_\omega) = \emptyset$.

Remark. Theorem 15 extends to the case where the assumptions on $(\alpha_i(\omega), i \in \mathbb{Z})$ to be i.i.d. random variables are relaxed to $\{\alpha_i, i \in \mathbb{Z}\}$ an ergodic (i.e. metrically transitive) stochastic process with $|\alpha_i(\omega)| \leq c$, $\forall \ i \in \mathbb{Z}, \omega \in \Omega$.
The other results extend to this case too, provided $\operatorname{supp} P = C^{\mathbb{Z}}$ with $C \in \operatorname{supp} P_{\alpha_o}$ compact (this is e.g. satisfied for α_i a stationary Markov process with strict positive transition function $p(x,A) > 0 \ \forall x \in C$, $\forall$ open $A \subset C$).

Let us now consider the stochastic Hamiltonian $-\Delta_{\alpha, Y(\omega)}$, with $Y(\omega) \equiv \{y_i + \xi_i(\omega)\}$, y_i being in $\Delta \equiv \{n\delta, n \in \mathbb{Z}\}$, and the (α_i, ξ_i) being i.i.d. We shall assume that in each unit cell $C_i \equiv \{x \in \mathbb{R} | i\delta - \frac{\delta}{2} \leq x \leq i\delta + \delta/2\}$ there exists one and only one center i.e. $\{\xi_i(\omega)| \leq \frac{\hat{\delta}}{2} < \frac{\delta}{2}$, so that

$$|x_i + \xi_i - (x_j + \xi_j)| \geq a - \hat{a} > 0 \text{ for all } i \neq j; \text{ for } j = i+1$$

$$|x_i + \xi_i - (x_j + \xi_j)| < a + \hat{a} \text{ for } j = i+1.$$

We then have:

Theorem. Let $\{\alpha_i, \xi_i, Y\}$ as above. Let $\tilde{a} \equiv \inf \text{supp } P_{\alpha_o}$, $\tilde{b} \equiv \sup \text{ supp } P_{\alpha_o}$. Assume $0 < \tilde{a} < \tilde{b} < \infty$ and $\text{supp } P_{\xi_o} = [-\frac{\hat{a}}{2}, \frac{\hat{a}}{2}]$, for some $\hat{a} < a$. Then the random Hamiltonian $-\Delta_{\alpha,Y(\omega)}$ has at least n gaps in its spectrum, provided $\hat{a} < a_n$ for some $a_n = a_n(\tilde{a}, \tilde{a})$.

Proof. See [7].

Remark. The assumption on i.i.d. can be relaxed, cfr. [47], [6] .

Remark. a) Most of the results of this section extend to the case where the point interaction $-\Delta_{\alpha,Y} = -\frac{d^2}{dx^2} + \sum \alpha_i \delta(x-y_i)$ is replaced by $-\frac{d^2}{dx^2} + \sum \alpha_i \delta'(x-y_i)$ with δ' the derivative of the δ-function. For a rigorous and detailed discussion of this case see [47], [11] .

b) A result on localization has been obtained in [19] , under the assumption of an absolutely continuous P_{α_o}.

Final remarks

a) We have not discussed the 2-dimensional case. Most results have been worked out also in this case, see [11] .

b) The case of interactions supported by 2-dimensional manifolds in $\mathbb{R}^3$ is discussed at least in the deterministic case, in [24] . A discussion, including the stochastic case, of the existence of the Hamiltonian for point interactions on random submanifolds of $\mathbb{R}^d$ is given in [25]. In particular the case of a heuristic Hamiltonian of the form $-\Delta + \lambda \int_0^t \delta(x-b(s))ds$ describing the interaction of a quantum mechanical particle in $\mathbb{R}^d$ interacting with a polymer in $\mathbb{R}^d$, modelled by the random paths b(s) of Brownian motion issued, say, at the origin, has been discussed in [25], [29] . This study has important connections with the statistical mechanics of polymers, see [25], [2C], [29] and quantum field theoretical models, see [25], [29] .

c) Recently a general study of Laplacian with boundary conditions on many small randomly distributed balls has been developed, see [48] and references therein. This theory has recently been put in connection with the study of random Hamiltonians with point interactions. Unfortunately we cannot go here into these fascinating developments and we have to refer the reader to [46], [50].

Acknowledgements

It is a great pleasure for the first author to thank Professor Erik Balslev and the Institute of Mathematics of Aarhus University for a kind invitation to a most interesting conference. The partial financial support of the Norwegian Research Council for Science and the Humanities and the Volkswagenstiftung (BiBoS-Project) as well as the Institute of Mathematics of Oslo University, the Institut de Physique Théorique of the Université Paris Sud (Orsay), the Centre de Physique Théorique, Université d'Aix-Marseille II, the Courant Institute of Mathematical Sciences is also gratefully acknowledged. We thank Mrs. Mischke and Mrs. Richter for their patience and skilful typing.

References

[1] A.M. Berthier, Spectral theory and wave operators for the Schrödinger equation, Pitman, London (1982)

[2] M.S.P. Eastham, The spectral theory of periodic differential equations, Scottish Academic Press, Edinburgh (1973)

[3] M. Reed, B. Simon, Methods of modern mathematical physics, Vol. IV Academic Press, New York (1978)

[4] M.M. Skriganov, The spectrum band structure of the three-dimensional Schrödinger operator with periodic potential, Inv. Math. 80, 107-121 (1985)

[5] B.E.J. Dahlberg, E. Trubowitz, A remark on two dimensional periodic potentials, Comment. Math. Helv. 57, 130-134 (1982)

[6] W. Kirsch, Über Spektren stochastischer Schrödinger Operatoren, Ph. D. Thesis, Bochum (1981)

[7] W. Kirsch, F. Martinelli, Some results on the spectra of random Schrödinger operators and their application to random point interaction models in one and three dimensions, pp. 223-244 in "Stochastic Processes in Quantum Theory and Statistical Physics", Proc. Marseille 1981, Edts. S. Albeverio,

Ph. Combe, M. Sirugue-Collin, Lect. Notes Phys. 173, Springer, Berlin (1985)

[8] R. Carmona, Random Schrödinger Operators, Ecole d'Eté de Probabilité XIV, Saint Flour, 1984

[9] E.H. Lieb, D.C. Mattis, Mathematical Physics in One Dimension, Academic Press, New York (1966)

[10] M. Gaudin, La fonction d'onde de Bethe, Masson, Paris (1983)

[11] S. Albeverio, F. Gesztesy, R. Høegh-Krohn, H. Holden, Solvable models in quantum mechanics, book in preparation

[12] Y.N. Demkov, V.N. Ostrovskii, The use of zero-range potentials in atomic physics (in russian), Nauka, Moscow (1975)

[13] S. Albeverio, Analytische Lösung eines idealisierten Stripping- oder Beugungsproblems, Helv. Phys. Acta 40, 135-184 (1967)

[14] A. Grossman, T.T. Wu, A class of potentials with extremely narrow resonances. I Case with discrete rotational symmetry, Preprint Marseille (1984)

[15] G.H. Wannier, Elements of solid state theory, Cambridge University Press (1959)

[16] S. Albeverio, R. Høegh-Krohn, Schrödinger operators with point interactions and short range expansion, Physica 124A, 11-28 (1984)

[17] S. Albeverio, F. Gesztesy, R. Høegh-Krohn, H. Holden, Some exactly solvable models in quantum mechanics and the low energy expansions, pp. in "Proceedings of the Second International Conference on Operator Algebras, Ideals, and Their Applications in Theoretical Physics, Leipzig 1983, Ed. H. Baumgärtel, G. Lassner, A. Pietsch, A. Uhlmann, Teubner, Leipzig (1984)

[18] Proceedings of the Meeting on "Random Media", Institute for Mathematics and its Applications, Minneapolis, Minnesota (1986)

[19] F. Delyon, B. Simon, B. Souillard, Ann. Inst. H. Poincaré 42, 283- (1985)

[20] W. Kirsch, F. Martinelli, On the spectrum of Schrödinger operators with a random potential, Comm. Math. Phys. 85, 329-350 (1982)

[21] H. Englisch, One-dimensional Schrödinger operators with ergodic potential, Zeitschr. f. Analys. und ihre Anw. 2, 411-426 (1983)

[22] H. Englisch, Instabilities and δ-distributions in solid state physics, Leipzig Preprint

[23] S. Albeverio, R. Høegh-Krohn, W. Kirsch, F. Martinelli, The spectrum of the three-dimensional Kronig-Penney model with random point defects, Adv. Appl. Math. 3, 435-440 (1982)

[24] J. Shabani, Modéles exactement solubles d'intéractions de surface en mécanique quantique non relativiste, Thése, Univ. de Louvain (1986)

[25] S. Albeverio, J.E. Fenstad, R. Høegh-Krohn, T. Lindstrøm, Nonstandard methods in stochastic analysis and mathematical physics, Academic Press, New York (1986)

[26] S. Albeverio, R. Høegh-Krohn, M. Mebkhout, Scattering by impurities in a solvable model of a 3-dimensional crystal, J. Math. Phys. 25, 1327-1334 (1984)

[27] E. Van Beveren, C. Dullenmond, T.A. Rijken, G. Rupp, An analytically solvable multichannel Schrödinger model for hadron spectroscopy, 182-191 in "Resonances - Models and Phenomena" Ed. S. Albeverio, L.S. Fereira, L. Streit, Lect. Notes Phys., Springer (1984)

[28] S.F. Edwards, Y.B. Gulyaev, Proc. Phys. Soc. 83, 495 (1964)

[29] S. Albeverio, J.E. Fenstad, R. Høegh-Krohn, W. Karwowski, T. Lindstrøm, Perturbations of the Laplacian supported by null sets, with applications to polymer measures and quantum fields, Phys. Letts. 104, 396-400, (1984)

[30] M. Reed, B. Simon, Methods of Modern Mathematical Physics II, Academic Press, New York (1975)

[31] E. Nelson, Internal set theory: a new approach to non standard analysis, Bull. Am. Math. Soc. 83, 1165-1198 (1977)

[32] S. Albeverio, J.E. Fenstad, R. Høegh-Krohn, Singular perturbations and nonstandard analysis, Trans. Am. Math. Soc. 252, 275-295 (1979)

[33] S. Albeverio, F. Gesztesy, R. Høegh-Krohn, The low energy expansion in nonrelativistic scattering theory, Ann. Inst. H. Poincaré Sect. A 37, 1-28 (1982)

[34] A. Grossmann, R. Høegh-Krohn, M. Mebkhout, The one particle theory of periodic point interactions, Comm. Math. Phys. 77, 87-110 (1980)

[35] S. Albeverio, R. Høegh-Krohn, L. Streit, Energy forms, Hamiltonians and distorted Brownian paths, J. Math. Phys. 18, 907-917 (1977)

[36] M. Fukushima, Energy forms and diffusion processes, pp. 65-98 in "Mathematics + Physics", Vol. 1, Ed. L. Streit, World Scient., Singapore (1985)

[37] S. Albeverio, R. Høegh-Krohn, Diffusion fields, quantum fields and fields with values in Lie groups, pp. 1-98 in M. Pinsky, Edt., Stochastic Analysis and Applications, M. Dekker, New York (1984)

[38] L. Dabrowski, H. Grosse, On nonlocal point interactions in one, two and three dimensions, J. Math. Phys. 26, 2777-2780 (1985)

[39] J. Brasche, Perturbations of Schrödinger Hamiltonians by measures - self - adjointness and lower semiboundedness, J. Math. Phys. 26, 621-626 (1985) and paper in preparation

[40] H. Holden, R. Høegh-Krohn, M. Mebkhout, The short-range expansion in solid state physics, Ann. Inst. H. Poincaré Sect. A 41, 333-360 (1984)

[41] J. Bellissard, R. Lima, D. Testard, Almost periodic Schrödinger operators, pp. 1-64 in "Mathematics + Physics", Vol. 1, Ed. L. Streit, World Scient. Singapore (1985)

[42] H. Holden, W. Kirsch, unpublished

[43] W. Kirsch, F. Martinelli, On the ergodic properties of the spectrum of general random operators, Jorn. Reine Angew. Math. 334, 141-158 (1982)

[44] S. Albeverio, F. Gesztesy, R. Høegh-Krohn, W. Kirsch, On point interactions in one dimension, J. Oper. Theory 12, 101-126 (1984)

[45] D. Testard, An explicitely solvable model of periodic point interaction in one dimension, pp. 371-383 in "Trends and Developments in the Eighties", Edts. S. Albeverio, Ph. Blanchard

[46] W. Kirsch, to appear

[47] F. Gesztesy, H. Holden, W. Kirsch, On Energy Gaps in a New Type of Analytically Solvable Model in Quantum Mechanics, Orsay-Preprint, 1986

[48] R. Figari, E. Orlandi, S. Teta, The Laplacian in regions with many small obstacles: Fluctuations around the limit operator, J. of Stat. Phys. Vol. 41 (1985) S. 465-487

[49] R. Figari, S. Teta, in preparation

[50] J.M. Luttinger, Wave propagation in one-dimensional structure, reprinted in [9]

[51] A. Grossmann, R. Høegh-Krohn, M. Mebkhout, A class of explicitely soluble, local, many-centers Hamiltonians for one-particle quantum mechanics in 2 and 3 dimensions, I, J. Math. Phys. 21, 2376-2385 (1980)

Wave operators for dilation-analytic three - body Hamiltonians

Erik Balslev

Matematisk Institut, Ny Munkegade
DK - 8000 Aarhus C, Denmark

Introduction.

In the present paper we give an exposition of some recent results on spectral projections and wave operators for dilated three - body Hamiltonians. The emphasis is on the main results and their proof. For several technical points omitted here we refer to the more detailed version given in [2]. A different approach to these problems, based on the Weinberg-van Winter equation, was introduced by C. van Winter [7].

We consider a three-body Schrödinger operator with dilation-analytic, multiplicative pair potentials decaying faster than r^{-2} at ∞. For such potentials the set of dilation-analytic resonances is generically finite inside any angle smaller than the angle of dilation-analyticity defined by the potential. This follows from a result on boundedness of resonances proved in [3] and simple analyticity considerations (Remarks 1.3 - 4). This situation is very suitable for the study of globally defined spectral projections and wave operators for small dilation angle φ and in particular for the investigation of the limit for $\varphi \to 0$, based on uniform estimates obtained in [3].

The approach is based on Kato's theory of smooth perturbations [6] and the abstract stationary scattering theory, utilizing the symmetrized Faddeev equation ([1],[8]).

In section 2 we quote some basic technical results derived in [2], on the boundary values of the various components of the resolvent, extending results of [6] to the present situation.

Utilizing these boundary values, we construct in section 3 a spectral measure for the dilated Hamiltonian. In section 4 we prove the existence and completeness of channel wave operators for each channel. Generically, there is a finite number of channels, and we have completeness for the dilated operator.

The question now arises about the connection of these dilated channel wave operators with the channel wave operators for the given, un-dilated Hamiltonian. In section 5 we prove that for all channels α, the inverse channel wave operator $F_{\alpha+}(\varphi,\Delta)$ converges strongly as $\varphi \to 0_+$ to the inverse channel wave operator $F_{\alpha+}(\Delta)$ for the given Hamiltonian. Here $\Delta \subseteqq [\lambda_\alpha, \lambda_\alpha']$, where λ_α is the threshold associated with the channel α and λ_α' is the next higher threshold. For $\alpha = 0$, $\lambda_\alpha' = \infty$. Moreover, for dilation-analytic vectors f, $F_{\alpha+}(\varphi,\Delta)f$ is the analytic continuation to $\{e^{-2i\varphi}\mathbb{R}^+\}$ of $F_{\alpha+}(\Delta)f$. (Theorem 5.2).

The operators $F_{\beta+}(\varphi,\Delta)$ corresponding to thresholds $\lambda_\beta < \lambda_\alpha$ have strong limits as $\varphi \to 0_+$, but they do not converge to $F_{\beta+}(\Delta)$ for $\Delta \subseteqq [\lambda_\alpha, \lambda_\alpha']$. This indicates that asymptotic completeness can not be proved by this method, although there is completeness for the dilated operators. Also the spectral projection operators $E_\alpha(\varphi,\Delta)$ have only been proved to have weak limits as $\varphi \to 0_+$ with no obvious interpretation of these limits.

1. Definitions and background.

We use the following standard notations for the three-body problem. Let particles 1,2,3 have masses m_1, m_2, m_3 and position vectors x_1, x_2, x_3 in $\mathbb{R}^3$. Denote pairs (ij) by α, β etc., and if $\alpha = (12)$, for example, let $m_\alpha^{-1} = m_1^{-1} + m_2^{-1}$, $n_\alpha^{-1} = (m_1 + m_2)^{-1} + m_3^{-1}$, $x_\alpha = x_2 - x_1$, $y_\alpha = x_3 - \frac{m_1 x_1 + m_2 x_2}{m_1 + m_2}$. The conjugate momenta are denoted by k_α, p_α.

For $\alpha \neq \beta$, the change of variables is given by

$$\begin{pmatrix} x_\beta \\ y_\beta \end{pmatrix} = \begin{Bmatrix} t_1 & t_2 \\ t_3 & t_4 \end{Bmatrix} \begin{pmatrix} x_\alpha \\ y_\alpha \end{pmatrix} \tag{1.1}$$

The basic Hilbert space is $H = L^2(\mathbb{R}^6)$, and $H^2(\mathbb{R}^6)$ is the Sobolev space of order 2 on $\mathbb{R}^6$. We introduce the following anxiliary Hilbert and Banach spaces, $s \in \mathbb{R}$,

$$L_s^2(\mathbb{R}^n) = \{g \mid \|g\|_{2,s}^2 = \int_{\mathbb{R}^n} |g(x)|^2 (1 + |x|^2)^s \, dx < \infty\}, \quad n = 3,6.$$

$$L_s^1(\mathbb{R}^3) = \{g \mid \|g\|_{1,s} = \int_{\mathbb{R}^3} |g(x)| (1 + |x|)^s \, dx < \infty\}.$$

$$L_s^\infty(\mathbb{R}^3) = \{g \mid \|g\|_{\infty,s} = \operatorname*{ess\,sup}_{x \in \mathbb{R}^3} |g(x)| (1 + |x|)^s < \infty\}.$$

$$M_s^P(\mathbb{R}^3) = L^P(\mathbb{R}^3) \cap L_s^1(\mathbb{R}^3) + L_s^\infty(\mathbb{R}^3).$$

$$H_\alpha = L^2(\mathbb{R}^3_{y_\alpha})_{y_\alpha}, \quad H_{\alpha,s} = L_s^2(\mathbb{R}^3_{y_\alpha})_{y_\alpha}.$$

$$\tilde{H} = \sum_\alpha {}_\oplus \{H \oplus H_\alpha\}, \quad \tilde{H}_{-s} = \sum_\alpha \{H \oplus H_{\alpha,-s}\}.$$

$$\overline{H} = H \oplus \sum_\alpha H_\alpha, \quad \overline{H}_s = L_s^2(\mathbb{R}^6) \oplus \sum_\alpha {}_\oplus H_{\alpha,s}.$$

$$X = \sum_\alpha \{L_s^2(\mathbb{R}^3_{x_\alpha}) \otimes L^2(\mathbb{R}^3_{y_\alpha})\}, \quad \overline{H}'_s = X \oplus \sum_\alpha {}_\oplus H_{\alpha,s}.$$

$$h_\alpha = L^2(S^2_{p_\alpha}), \quad \text{where} \quad S^2_{p_\alpha} = \{p_\alpha \in \mathbb{R}^3 \mid |p_\alpha| = 1\}$$

$$h_0 = L^2(S^5), \quad \text{where} \quad S^5 = \{(k_\alpha, p_\alpha) \mid |(k_\alpha, p_\alpha)| = 1\}.$$

The weight function f_s is defined by

$$f_s(x) = (1+|x|)^{-s} , \; x \in \mathbb{R}^3 .$$

The free Hamiltonian H_0 is defined for any α by

$$H_0 = -\frac{\Delta_{x_\alpha}}{2m_\alpha} - \frac{\Delta_{y_\alpha}}{2n_\alpha} , \quad \mathcal{D}(H_0) = H^2(\mathbb{R}^6) .$$

Let $\{U(\rho) \mid \rho \in \mathbb{R}^+\}$ be the dilation - group on $\mathbb{R}^3$, and let $V(\rho) = U(\rho)\, V\, U(\rho^{-1})$. Let $S_a = \{\rho e^{i\varphi} \mid \rho > 0 , \; |\varphi| < a\}$, where $a \leqq \frac{\pi}{2}$. The interaction is assumed to be a sum of pair potentials V_α, each of which satisfies the following condition

A. V is a Δ - compact, multiplicative, real potential, and $V(\rho) : \mathbb{R}^+ \to C(\mathcal{D}_\Delta , H)$ has an analytic, $C(\mathcal{D}_\Delta , H)$ - valued extension $V(\rho e^{i\varphi})$ to S_a. For each $\varphi \in (-a,a)$, $V(\varphi) = V(e^{i\varphi}) \in M^p_{2s}(\mathbb{R}^3)$ for some $p > \frac{3}{2}$, $s > 1$.

We consider the self - adjoint, analytic families of type A defined by

$$h_\alpha(z) = -z^{-2}\Delta_{x_\alpha} + V_\alpha(z) , \quad \mathcal{D}(h_\alpha(z)) = H^2(\mathbb{R}^3_{x_\alpha}) , \; z \in S_a$$

$$H(z) = z^{-2} H_0 + \sum_\alpha V_\alpha(z) , \quad \mathcal{D}(H(z)) = H^2(\mathbb{R}^6) , \; z \in S_a .$$

We consider first the two - body operators $h_\alpha(z)$. The essential spectrum of $h_\alpha(z)$ is $e^{-2i\varphi}\,\overline{\mathbb{R}^+}$. We denote by $R_\alpha(\varphi)$ the non - real discrete spectrum of $h_\alpha(z)$ and by $R_\alpha = \bigcup_{0<\varphi<a} R_\alpha(\varphi)$ the set of resonances of h_α. Some known results on the operators $h_\alpha(\varphi)$ are stated in the following lemma.

Lemma 1.1. (i) The discrete spectrum of h_α is finite.

(ii) Eigenfunctions ϕ_α corresponding to the eigenvalue $\lambda_\alpha < 0$

satisfy for some constants $b,c > 0$

$$|\phi_\alpha(x)| \leqq c\, e^{-b|x|} \quad , \; x \in \mathbb{R}^3 \; .$$

(iii) $R_\alpha(\varphi)$ is bounded for every $\varphi \in (0,a)$. Also the set of positive eigenvalues of h_α is bounded and accumulates at most at 0 .

(iv) For $0 \leqq \varphi < a$, $\mu \geqq 0$, the following limits exist in the operator norm topology of $B(H)$,

$$B_\alpha(\varphi)(-\Delta_{x_\alpha} - \mu)_+^{-1} A_\alpha(\varphi) = \lim_{\varepsilon \downarrow 0} B_\alpha(\varphi)(-\Delta_{x_\alpha} - \mu - i\varepsilon)^{-1} A_\alpha(\varphi) \; ,$$

where $V_\alpha(\varphi) = A_\alpha(\varphi) B_\alpha(\varphi)$, $B_\alpha(\varphi) = |V_\alpha(\varphi)|^{\frac{1}{2}}$.

In view of Lemma 1.1 (i) the following assumption is essentially a simplification of notation.

B. For each α , h_α has one simple, negative eigenvalue with corresponding normalized eigenfunction ϕ_α .
To simplify notations, we also allow α to take the value 0 and set $\lambda_0 = 0$. Using Lemma 1.1 (iv), we introduce the further assumption

C. h_α has no zero - energy resonance or eigenvalue for any α , i.e.

$$N(1 + B_\alpha(-\Delta_{x_\alpha})^{-1} A_\alpha) = \{0\} \qquad \text{for all } \alpha \; .$$

where $B_\alpha(-\Delta_{x_\alpha})^{-1} A_\alpha = \lim_{\varepsilon \downarrow 0} B_\alpha(-\Delta_{x_\alpha} - i\varepsilon)^{-1} A_\alpha$ in $B(L^2_{x_\alpha})$.

<u>Remark 1.2</u>. $K_\alpha(z) = B_\alpha(z)(-\Delta_{x_\alpha} z^{-2})^{-1} A_\alpha(z)$ is an analytic, $C(L^2_{x_\alpha})$ - valued function of $z \in S_a$, such that

$$K_\alpha(\rho e^{i\varphi}) = U_\alpha(\rho) K_\alpha(e^{i\varphi}) U_\alpha(\rho^{-1}) \quad \text{for} \quad \varphi \in (-a,a) \; .$$

Hence, C implies that

$$N(1 + B_\alpha(z)(-\Delta_{x_\alpha} z^{-2})^{-1} A_\alpha(z) = \{0\} \quad \text{for all } \alpha \text{ and } z \in S_a \; .$$

Remark 1.3. It follows from C, that resonances and positive eigenvalues do not accumulate at 0. In view of Lemma 1.1 (iii) this implies that the number of positive eigenvalues and the set of resonances $R_{a-\varepsilon}$ is finite for $\varepsilon > 0$.

Remark 1.4. C is satisfied generically, i.e. except for a sequence $\{\kappa_n\}$ of coupling constants, where $|\kappa_n| \underset{n\to\infty}{\to} \infty$, the operators $-\Delta_{x_\alpha} + \kappa V_\alpha$ satisfy C for all α. This follows from the analyticity of the $C(L^2_{x_\alpha})$ - valued function $\kappa K_\alpha(1)$ by the analytic Fredholm theorem.

The following assumption will also be needed:

D. For a given $\varphi \in (0,a)$, $e^{-2i\varphi}\mathbb{R}^+$ contains no two - body resonances, i.e. $e^{-2i\varphi}\mathbb{R}^+ \cap R_\alpha(\varphi+\delta) = \emptyset$ for $\delta > 0$ and all α.

Remark 1.5. In view of Remark 1.3, D is satisfied for all $\varphi \in (0,a-\varepsilon]$, $\varepsilon > 0$, except at most a finite number of values.

Conditions A - D are standing assumptions throughout the paper.

Later on we shall need the following condition:

E.
$$V \in L^{2p}_s(\mathbb{R}^3) \cap L^2_{2s}(\mathbb{R}^3) + L^\infty_{2s}(\mathbb{R}^3)$$

for some $p > \frac{3}{2}$ and $s > 1$.

We turn now to the three - body problem.

The essential spectrum of $H(z)$ is $\underset{\lambda}{\cup}\{\lambda + e^{-2i\varphi}\overline{\mathbb{R}^+}\}$ where λ ranges over 0 and all discrete eigenvalues and resonances of the operators h_α. The non - real, discrete spectrum of $H(z)$, denoted by $R(\varphi)$, is located between the half - lines $\{\lambda_e + \mathbb{R}^+\}$ and $\{\lambda_e + e^{-2i\varphi}\mathbb{R}^+\}$, where $\lambda_e = \min_\alpha \lambda_\alpha$, and otherwise φ - independent, unless "absorbed" by the essential spectrum.

The self-adjoint analytic family $H_\alpha(z)$ of operators is defined for $z \in S_a$ by

$$H_\alpha(z) = z^{-2} H_0 + V_\alpha(z) \quad , \quad \mathcal{D}_{H_\alpha(z)} = H^2(\mathbb{R}^6) \ .$$

The resolvent $R_\alpha(z,\zeta)$ can be decomposed as

$$R_\alpha(z,\zeta) = R^0_\alpha(z,\zeta) + |\phi_\alpha(z)\rangle \tilde{r}_\alpha(z,\zeta) \langle \bar{\phi}_\alpha(z)|$$

where

$$\tilde{r}_\alpha(z,\zeta) = \left(- z^{-2} \frac{\Delta y_\alpha}{2n_\alpha} + \lambda - \zeta\right)^{-1}$$

and

$$(\langle \chi | g)(y_\alpha) = \int_{\mathbb{R}^3_{x_\alpha}} \overline{\chi(x_\alpha)} g(x_\alpha, y_\alpha) dx_\alpha \quad \text{for} \quad g \in H, \chi \in L^2(\mathbb{R}^3_{x_\alpha})$$

$$(|\chi\rangle \sigma)(x_\alpha, y_\alpha) = \chi(x_\alpha)\sigma(y_\alpha) \qquad \text{for} \quad \chi \in L^2(\mathbb{R}^3_{x_\alpha}), \sigma \in H_\alpha$$

The Faddeev matrix $A(z,\zeta)$ is defined as follows:

$$Q_{\alpha\beta}(z,\zeta) = \begin{cases} B_\alpha(z) R^0_\beta(z,\zeta) A_\beta(z) & B_\alpha(z) |\phi_\beta(z)\rangle \\ \\ \tilde{r}_\alpha(z,\zeta) \langle \bar{\phi}_\alpha(z)| V_\alpha(z) R^0_\beta(z,\zeta) A_\beta(z) & \tilde{r}_\alpha(z,\zeta) \langle \bar{\phi}_\alpha(z)| V_\alpha(z) |\phi_\beta(z)\rangle \end{cases}$$

$$A(z,\zeta) = \begin{cases} 0 & Q_{\alpha\beta}(z,\zeta) & Q_{\alpha\gamma}(z,\zeta) \\ Q_{\beta\alpha}(z,\zeta) & 0 & Q_{\beta\alpha}(z,\zeta) \\ Q_{\gamma\alpha}(z,\zeta) & Q_{\alpha\beta}(z,\zeta) & 0 \end{cases}$$

with $\alpha = (12)$, $\beta = (23)$, $\gamma = (31)$.

The operators $H(z,\zeta) \in \mathcal{B}(H,\tilde{H})$ are defined for $\zeta \in \sigma_e(H(z))$ by

$$H(z,\zeta) = \{\theta_{\alpha,0}H(z,\zeta) \ , \ \theta_{\alpha,1}H(z,\zeta)\}_{\alpha = (12),(23),(31)}$$

where

$$\theta_{\alpha,0}H(z,\zeta) = \sum_{\beta \neq \alpha} B_\alpha(z) R_0(z,\zeta) V_\beta(z) R_\beta(z,\zeta)$$

$$\theta_{\alpha,1}H(z,\zeta) = \tilde{r}_\alpha(z,\zeta) \langle \bar{\phi}_\alpha(z)| A_\alpha(z) \theta_{\alpha,0}H(z,\zeta) \ .$$

Moreover $K(z,\zeta) \in \mathcal{B}(H,\tilde{H})$ is defined by

$$K(z,\zeta) = (1 + A(z,\zeta))^{-1} H(z,\zeta) \ .$$

It is proved in [2] that $(1+A(z,\zeta))^{-1}$ is meromorphic in $\mathbb{C}\setminus\sigma_e(H(z))$ with poles at points of $R(\varphi)$. Hence $K(z,\zeta)$ is meromorphic in the same region with poles contained in $R(\varphi)$.

The operators $Y(z,\zeta) \in B(H,\overline{H})$ are defined by

$$Y(z,\zeta) = \{Y_\alpha(z,\zeta)\}, \quad \alpha = 0, (12), (23), (31)$$

where

$$Y_0(z,\zeta) = -2I + \sum_\alpha (1 - |\phi_\alpha(z)\rangle\langle\overline{\phi}_\alpha(z)|) + Y_0'(z,\zeta)$$

with

$$Y_0'(z,\zeta) = \sum_\alpha (1 - |\phi_\alpha(z)\rangle\langle\overline{\phi}_\alpha(z)|) A_\alpha(z)\theta_{\alpha,0}K(z,\zeta)$$
$$-\sum_\alpha V_\alpha(z) R_\alpha^0(z,\zeta)(1 + A_\alpha(z)\theta_{\alpha,0}K(z,\zeta)$$

and

$$Y_\alpha(z,\zeta) = \langle\overline{\phi}_\alpha(z)| + Y_\alpha'(z,\zeta)$$

with

$$Y_\alpha'(z,\zeta) = \langle\overline{\phi}_\alpha(z)| A_\alpha(z)\theta_{\alpha,0}K(z,\zeta).$$

The identification operator $J(z) \in B(\overline{H},H)$ is given by

$$J(z)(u,\tau_{12},\tau_{23},\tau_{31}) = u + \sum_\alpha |\phi_\alpha(z)\rangle\tau_\alpha .$$

Let $R_1(z,\zeta) = R_0(z,\zeta) \oplus \sum_{\alpha\oplus} \tilde{r}_\alpha(z,\zeta)$.

The following decomposition formula is proved in [8], see also [2].

$$R(z,\zeta) = J(z)R_1(z,\zeta)Y(z,\zeta). \tag{1.2}$$

The operator $G(z,\zeta) \in B(\overline{H},H)$ is defined by

$$G(z,\zeta)(u,\tau_\alpha,\tau_\beta,\tau_\gamma) = G_0(z,\zeta)u + \sum_\alpha G_\alpha(z,\zeta)\tau_\alpha$$

where

$$G_0(z,\zeta)u = (1 + V(z)R_0(z,\zeta))u$$
$$G_\alpha(z,\zeta)\tau_\alpha = (|\phi_\alpha(z)\rangle + W_\alpha(z)\tilde{r}_\alpha(z,\zeta))\tau_\alpha$$

and

$$W_\alpha(z) = \sum_{\beta\neq\alpha} V_\beta(z)|\phi_\alpha(z)\rangle .$$

The trace operators are defined as follows.

Let F_α be the Fourier transform on $L^2(\mathbb{R}^3_{y_\alpha})$ and let F_0 be the Fourier transform on $L^2(\mathbb{R}^6)$. For $f \in L^2_s(\mathbb{R}^3_{y_\alpha})$, $g \in X$, set

$$\gamma_\alpha(\mu)f = (F_\alpha f)(\mu,\cdot) , \quad \mu \in \mathbb{R}^+$$

$$\gamma_0(\mu)g = (F_0 g)(\mu,\cdot) , \quad \mu \in \mathbb{R}^+$$

where we use polar coordinates in $\mathbb{R}^3$ and $\mathbb{R}^6$ respectively. The following result is well known (cf.[4] Proposition (2.1) and (2.2)).

<u>Lemma 1.6</u>. For $1 < s < \frac{3}{2}$, $\mu \in \mathbb{R}^+$ there exists $C < \infty$ such that

$$\gamma_\alpha(\mu) \in B(L^2_s(\mathbb{R}^3),h_\alpha) , \quad \|\gamma_\alpha(\mu)\| < C\min\{\mu^{-1},\mu^{s-\frac{3}{2}}\} \tag{1.3}$$

$$\gamma_0(\mu) \in B(X, h_0) , \quad \|\gamma_0(\mu)\| < C\min\{\mu^{-2},\mu^{s-3}\} \tag{1.4}$$

Set

$$c_\alpha(\mu) = (2n_\alpha)^{3/4}\,\frac{1}{2}\,\mu^{1/4} , \quad c_0(\mu) = (2m)^{3/2}\,\frac{1}{2}\,\mu$$

$$T_\alpha(\mu) = c_\alpha(\mu)\gamma_\alpha((2n_\alpha\mu)^{\frac{1}{2}})$$

$$T_0(\mu) = c_0(\mu)\gamma_0((2m\,\mu)^{\frac{1}{2}}) .$$

2. Boundary values on the essential spectrum.

In this section we collect the basic results on the existence of boundary values of the operators $A(\varphi,\lambda_\alpha + e^{-2i\varphi}(\mu \pm i\varepsilon))$ and $Y_\alpha(\varphi,\lambda_\alpha + e^{-2i\varphi}(\mu \pm i\varepsilon))$ as $\varepsilon \downarrow 0$.

Lemma 2.1. The following limits exist in the operator norm topology for all α, $0 \leqq \varphi < a$ and $\mu \geqq 0$,

$$A_\pm(\varphi,\lambda_\alpha + e^{-2i\varphi}\mu) = \lim_{\varepsilon \downarrow 0} A(\varphi,\lambda_\alpha + e^{-2i\varphi}(\mu \pm i\varepsilon)) \in B(\tilde{\mathcal{H}}_{-s}) .$$

Moreover,

$$A_\pm^2(\varphi,\lambda_\alpha + e^{-2i\varphi}\mu) \in C(\tilde{\mathcal{H}}_{-s}) .$$

Proof. We refer to [4] or [8].

Lemma 2.2. Let $-2\varphi < \operatorname{Arg}(\zeta - \lambda_\alpha) < 0$. Then ζ is a resonance if and only if

$$N(1 + A(\varphi,\zeta)) \neq \{0\} .$$

Let $\mu > 0$. Then $\lambda_\alpha + e^{-2i(\varphi \pm \varepsilon)}\mu$ is a resonance for $0 < \varepsilon < \varepsilon_0$ if and only if

$$N(1 + A_\pm(\varphi,\lambda_\alpha + e^{-2i\varphi}\mu)) \neq \{0\} .$$

Proof. We refer to [1].

Definition 2.3. λ_α is said to be a threshold resonance if

$$N(1 + A(\lambda_\alpha)) \neq \{0\}$$

where

$$A(\lambda_\alpha) = A_\pm(\lambda_\alpha) \qquad (\mu = 0) .$$

Note that we include the possibility that λ_α may be an eigenvalue of H, since this distinction is not relevant here.

Lemma 2.4. If $N(1 + A(\varphi_0,\lambda_\alpha)) \neq \{0\}$ for one $\varphi_0 \in (-a,a)$, then $N(1 + A(\varphi,\lambda_\alpha)) \neq \{0\}$ for all $\varphi \in (-a,a)$.

Proof. Take $\varphi_0 = 0$, $\varphi_0 \neq 0$ is similar. The Fredholm alternative holds for $A(z,\lambda_\alpha)$, since $A^2(z,\lambda_\alpha)$ is compact (cf.[4]). Hence the spectrum of $A(z,\lambda_\alpha)$ consists of a sequence of finite-dimensional eigenvalues $\mu_n(z)$ such that $\mu_n(z) \xrightarrow[n\to\infty]{} 0$. We have

$$A(\rho\, e^{i\varphi},\lambda_\alpha) = \tilde{U}(\rho) A(e^{i\varphi},\lambda_\alpha)\tilde{U}^{-1}(\rho)$$

where

$$\tilde{U}(\rho) = \sum_\alpha \oplus (U(\rho) \oplus U_\alpha(\rho)) .$$

If

$$A(\rho_1\, e^{i\varphi}, \lambda_\alpha)\phi + \mu_n \phi = 0 ,$$

then

$$A(\rho_2\, e^{i\varphi}, \lambda_\alpha)\tilde{U}(\rho_2\rho_1^{-1})\phi + \mu_n \tilde{U}(\rho_2\rho_1^{-1})\phi = 0$$

hence

$$\sigma(A(\rho\, e^{i\varphi}, \lambda_\alpha)) = \sigma(A(e^{i\varphi},\lambda_\alpha)) \quad \text{for all} \quad \rho > 0 .$$

Note that it is immaterial for this argument that $\tilde{U}(\rho)$ is not a unitary operator on $\tilde{H}_{-s}$.

Moreover, $A(z,\lambda_\alpha)$ is an analytic $B(\tilde{H}_{-s})$ - valued function of $z \in S_a$, hence the eigenvalues are (branches of) analytic functions of z . Since eigenvalues $\mu_n(z)$ of $A(z,\lambda_\alpha)$ are constant on rays $\{z = \rho\, e^{i\varphi_0} \mid \rho > 0\}$, a connectedness argument shows that each $\mu_n(z)$ is a constant μ_n for $z \in S_a$. In particular, if -1 is an eigenvalue of $A(\lambda_\alpha) = A(1,\lambda_\alpha)$, then -1 is an eigenvalue of $A(z,\lambda_\alpha)$ for all $z \in S_a$.

Lemma 2.5. "Generically", $N(1 + A(\lambda_\alpha)) = \{0\}$, i.e. if we introduce a coupling constant $\kappa \in \mathbb{C}$, setting $H(\kappa) = H_0 + \kappa V$, then there exists a sequence $\{\kappa_n\}$, $|\kappa_n| \xrightarrow[n\to\infty]{} \infty$, such that for $\kappa \neq \kappa_n$

$$N(1 + A(\lambda_\alpha,\kappa)) = \{0\} .$$

Proof. $A(\lambda_\alpha,\kappa)$ is an analytic $B(\tilde{H}_{-s})$ - valued function of $\kappa \in \mathbb{C}$. Since $A^2(\lambda_\alpha,\kappa) \in C(\tilde{H}_{-s})$, the function

$$(1 + A(\lambda_\alpha,\kappa))^{-1} = (1 - A(\lambda_\alpha,\kappa))(1 - A^2(\lambda_\alpha,\kappa))^{-1}$$

is meromorphic in $\mathbb{C}$ by the analytic Fredholm theorem. This implies the Lemma.

Lemma 2.6. If $N(1+A(\lambda_\alpha)) = \{0\}$, then for any $\varphi \in (0,a)$ the set of resonances R_φ does not accumulate at λ_α.

Proof. By Lemma 2.4, $N(1+A(\varphi,\lambda_\alpha)) = \{0\}$. Let the $B(\tilde{H}_{-s})$-valued function $\tilde{A}(\varphi,\zeta)$ be defined by

$$\tilde{A}(\varphi,\zeta) = \begin{cases} 1+A(\varphi,\zeta) & \text{for } \zeta = \lambda_\alpha \\ 1+A_+(\varphi,\zeta) & \text{for } \zeta = \lambda_\alpha + e^{-2i\varphi}\mu\,,\ \mu > 0 \\ 1+A(\varphi,\zeta) & \text{for } -2\varphi < \operatorname{Arg}(\zeta-\lambda_\alpha) < 0 \end{cases}$$

$\tilde{A}(\varphi,\zeta)$ is continuous at λ_α, hence $N(1+\tilde{A}(\varphi,\zeta)) = \{0\}$ for ζ near λ_α, and the Lemma is proved using Lemma 2.2.

Lemma 2.7. If $N(1+A(\lambda_\alpha)) = \{0\}$ for all α, then for any $\varphi \in (0,a)$ the set of resonances R_φ is finite.

Proof. This follows from Lemma 2.6 and the fact, proved in [3] that R_φ is bounded.

Corollary 2.8. Generically, the set of resonances R_φ is finite.

Basic for our approach is the following result, obtained by extending Kato's theory of smooth perturbations [6] to the operators utilized in the decomposition (1.2) of the resolvent.

Lemma 2.9. Assume that $\{\lambda_\alpha + e^{-2i\varphi}\overline{\mathbb{R}^+}\}$ contains no three-body resonances and λ_α is not a threshold resonance, $\alpha \neq 0$. Then the following limits exist in $L^2(\mathbb{R},H_{\alpha,s})$ for every $f \in H$,

$$Y'_{\alpha\pm}(\varphi,\lambda_\alpha + e^{-2i\varphi}\mu)f = \lim_{\varepsilon\downarrow 0} Y'_\alpha(\varphi,\lambda_\alpha + e^{-2i\varphi}(\mu\pm i\varepsilon))f \in L^2(\mathbb{R},H_{\alpha,s})\,.$$

For $\alpha = 0$ the following limits exist in $L^2(\mathbb{R},X)$ for every $f \in H$,

$$Y'_{0\pm}(\varphi, e^{-2i\varphi}\mu) f = \lim_{\varepsilon \downarrow 0} Y'_0(\varphi, e^{-2i\varphi}(\mu \pm i\varepsilon)) f \in L^2(\mathbb{R}, X) .$$

Proof. This follows from [2] Lemmas 2.15 and 2.17 and the theory of vector-valued Hardy functions (cf.[5]).

3. Spectral projections.

Theorem 3.1. Let $0 < \varphi < a$ and let $\delta \in \{(12),(23),(31),0\}$. Assume that $\{\lambda_\delta + e^{-2i\varphi}\mathbb{R}^+\}$ contains no three-body resonances, that λ_δ is not a threshold resonance and that $e^{-2i\varphi}\overline{\mathbb{R}^+}$ contains no two-body resonances or zero-energy resonance for any h_α. For $\delta = 0$ assume condition E.

For all $f, g \in H$ and any Borel set $\Delta \subset \mathbb{R}$ the following limits exist,

$$F_{\delta\varphi}(f,g;\Delta) = \lim_{\varepsilon \downarrow 0} F^{\varepsilon}_{\delta\varphi}(f,g;\Delta)$$

where

$$F^{\varepsilon}_{\delta\varphi}(f,g;\Delta) = \frac{1}{2\pi i} \int_\Delta (f, \{R(\varphi, \lambda_\delta + e^{-2i\varphi}(\mu+i\varepsilon)) - R(\varphi, \lambda_\delta + e^{-2i\varphi}(\mu - i\varepsilon))\} g)\, d\mu \tag{3.1}$$

There exist constants $C(\varphi) < \infty$ such that for all Borel sets $\Delta \subseteq \mathbb{R}$ and all $f, g \in H$

$$|F_{\delta\varphi}(f,g;\Delta)| \leq C(\varphi)\, \|f\| \cdot \|g\| . \tag{3.2}$$

Proof. We consider the case $\delta \neq 0$, say $\delta = \alpha = (12)$; the case $\delta = 0$ is similar. By the 1st resolvent equation and (1.2)

$$F^{\varepsilon}_{\alpha\varphi}(f,g;\Delta) = e^{-2i\varphi}\frac{\varepsilon}{\pi} \int_\Delta (R(-\varphi, \lambda_\alpha + e^{2i\varphi}(\mu+i\varepsilon)) f, R(\varphi, \lambda_\alpha + e^{-2i\varphi}(\mu+i\varepsilon) g)\, d\mu$$

$$= e^{-2i\varphi}\frac{\varepsilon}{\pi} \int_\Delta J(-\varphi) R_1(-\varphi, \lambda_\alpha + e^{2i\varphi}(\mu+i\varepsilon))\, Y(-\varphi, \lambda_\alpha + e^{2i\varphi}(\mu+i\varepsilon)) f ,$$

$$J(\varphi)R_1(\varphi,\lambda_\alpha + e^{-2i\varphi}(\mu+i\varepsilon))Y(\varphi,\lambda_\alpha + e^{-2i\varphi}(\mu+i\varepsilon))g)d\mu$$

$$= e^{-2i\varphi}\frac{\varepsilon}{\pi}\int_\Delta (\tilde{r}_\alpha(-\varphi,\lambda_\alpha + e^{2i\varphi}(\mu+i\varepsilon))Y_\alpha(-\varphi,\lambda_\alpha + e^{2i\varphi}(\mu+i\varepsilon))f\,,$$

$$\tilde{r}_\alpha(\varphi,\lambda_\alpha + e^{-2i\varphi}(\mu+i\varepsilon))Y_\alpha(\varphi,\lambda_\alpha + e^{-2i\varphi}(\mu+i\varepsilon))g)d\mu + O(\varepsilon) \qquad (3.3)$$

since all the remaining terms go to 0 as $\varepsilon \downarrow 0$ by [1b] Lemma 7.2.

We further obtain from (3.3) by the 1st resolvent equation and Lemmas 2.9 and 1.6

$$F^{\varepsilon}_{\alpha\varphi}(f,g;\Delta) \xrightarrow[\varepsilon \downarrow 0]{} e^{2i\varphi}\int_\Delta T_\alpha(\mu)Y_{\alpha+}(-\varphi,\lambda_\alpha + e^{2i\varphi}\mu)f\,,$$

$$T_\alpha(\mu)Y_{\alpha+}(\varphi,\lambda_\alpha + e^{-2i\varphi}\mu)g)_{h_\alpha}d\mu \qquad (3.4)$$

From Lemmas 2.9 and 1.6 we further obtain

$$\int_{\mathbb{R}} \|T_\alpha(\mu)Y_{\alpha+}(\varphi,\lambda_\alpha + e^{-2i\varphi}\mu)f\|^2_{h_\alpha} \leqq C(\varphi)\,\|f\|^2_H \qquad (3.5)$$

From (3.5) we obtain (3.2) by Schwarz' inequality, and the Theorem is proved.

Under the conditions of Theorem 3.1 we can now define the operators $E_\delta(\varphi,\Delta)$ by

$$(f,E_\delta(\varphi,\Delta)g) = e^{-2i\varphi}\,F_{\delta\varphi}(f,g;\Delta) \qquad (3.6)$$

Theorem 3.2. The operators $E_\delta(\varphi,\Delta)$ are uniformly bounded, i.e. $\|E_\delta(\varphi,\Delta)\| \leqq C(\varphi) < \infty$ for all Borel sets $\Delta \subseteqq \mathbb{R}$, and satisfy the following conditions:

(i) $E_\delta(\varphi,\Delta_1)E_\delta(\varphi,\Delta_2) = E_\delta(\varphi,\Delta_1 \cap \Delta_2)$

(ii) $E_\delta(\varphi,\Delta_1)E_\eta(\varphi,\Delta_2) = 0$ for $\delta \neq \eta$

(iii) $E_\delta(\varphi, \bigcup_{i=1}^\infty \Delta_i)f = \sum_{i=1}^\infty E_\delta(\varphi,\Delta_i)$ for $f \in H$ and $\Delta_i \cap \Delta_j = \emptyset, i \neq j$

(iv) $E_\delta(\varphi,\Delta)H(\varphi) \subseteqq H(\varphi)E_\delta(\varphi,\Delta)$.

Proof. The uniform boundedness of the operators $E_\delta(\varphi,\Delta)$ follows from (3.2). Property (iii) follows from the countable additivity of the sesquilinear forms,

$$F_{\delta\varphi}(f,g;\bigcup_{i=1}^{\infty}\Delta_1) = \sum_{i=1}^{\infty} F_{\delta\varphi}(f,g;\Delta_i)$$

which in turn follows from the expression for $F_{\delta\varphi}(f,g;\Delta)$ given by the right hand side of (3.4). Property (iv) follows from the expression of $F_{\delta\varphi}(f,g;\Delta)$ as the limit of the right hand side of (3.1) as $\varepsilon \downarrow 0$. Finally, (i) and (ii) are proved in the same way as [2b] Lemma 7.6.

Corollary 3.3. Assume that none of the thresholds are three-body resonances or eigenvalues, that $\{\lambda_\delta + e^{-2i\varphi}\mathbb{R}^+\}$ contains no three-body resonances for any δ and $\{e^{-2i\varphi}\overline{\mathbb{R}^+}\}$ contains no two-body resonances or eigenvalues and suppose that condition E is satisfied. Let $\{\lambda_1 \ldots \lambda_k\}$ be the set of discrete eigenvalues of $H(\varphi)$, and let $E_i(\varphi,\lambda_i)$ be the projection defined by

$$E_i(\varphi,\lambda_i) = \frac{-1}{2\pi i}\int_{\Gamma_i} R(\varphi,\xi)d\xi$$

where Γ_i is a circle centered at λ_i, separating λ_i from the rest of the spectrum of $H(\varphi)$.

Then $H(\varphi)$ is a spectral operator with the family of spectral projections $E(\varphi,\Omega)$ defined for Borel sets $\Omega \subset \mathbb{R}^2$ by

$$E(\varphi,\Omega) = \sum_\delta E(\varphi,\Omega \cap \{\lambda_\delta + e^{-2i\varphi}\mathbb{R}^+\}) + \sum_{\lambda_i \in \Omega} E_i(\varphi,\lambda_i) .$$

Remark 3.4. The assumption of absence of two- and three-body threshold resonances and eigenvalues is satisfied generically, i.e. except for a discrete set of coupling constants $\{\kappa_n\}$, $|\kappa_n| \xrightarrow[n\to\infty]{} \infty$. For a given $\kappa \notin \{\kappa_n\}$, the number of discrete eigenvalues of $H(\varphi)$

is finite, and except for a finite number of values of $\varphi \in (0,a-\varepsilon]$, $\{\lambda_\delta + e^{-2i\varphi}\mathbb{R}^+\}$ contains no three-body resonances and $\{e^{-2i\varphi}\mathbb{R}^+\}$ contains no two-body resonances.

4. Wave operators.

Under the assumptions of Theorem 3.1 we define for $0 < \varphi < a$ and Δ any Borel subset of $\mathbb{R}$ the operators $F_{\alpha\pm}(\varphi,\Delta) \in B(H,H_\alpha)$, $\alpha \in \{(12),(23),(31)\}$, and $F_{0\pm}(\varphi,\Delta) \in B(H)$ by setting for $f \in H$, $\delta = \alpha, 0$, $\mu \in \mathbb{R}^+$

$$(F_{\delta\pm}(\varphi,\Delta)f)(\mu) = \chi_\Delta(\mu) T_\delta(\mu) Y_{\delta\pm}(\varphi,\lambda_\alpha + e^{-2i\varphi}\mu) f \tag{4.1}$$

Here we use polar coordinates in $L^2(\mathbb{R}^3_{p_\alpha})$ and in $L^2(\mathbb{R}^6_p)$ for $\delta = 0$, and we utilize Lemmas 2.9 and 1.6.

Lemma 4.1. For all δ and Δ, the operator $F_{\delta\pm}(\varphi,\Delta)$ is bounded and one-to-one from $E_\delta(\varphi,\Delta)H$ into $L^2(\Delta,h_\delta)$ with a bounded inverse $F^{-1}_{\delta\pm}(\varphi,\Delta)$ satisfying

$$\|F^{-1}_{\delta\pm}(\varphi,\Delta)\| \leqq \|F_{\delta\pm}(\varphi,\Delta)\| \tag{4.2}$$

Proof. Boundedness follows from Lemmas 2.9 and 1.6. By the proof of Theorem 3.1

$$(f,E_\delta(\varphi,\Delta)g) = e^{2i\varphi}(F_{\delta+}(-\varphi,\Delta)f\,,\,F_{\delta+}(\varphi,\Delta)g)_{L^2(\Delta,h_\delta)} \tag{4.3}$$

hence

$$\|E_\delta(\varphi,\Delta)g\| \leqq \|F_{\delta+}(-\varphi,\Delta)\|_{L^2(\Delta,h_\delta)} \cdot \|F_{\delta+}(\varphi,\Delta)g\|_{L^2(\Delta,h_\delta)} \tag{4.4}$$

and the Lemma follows.

Lemma 4.2. For all δ and Δ, if $f \in H$ and $E_\delta(\varphi,\Delta)f = 0$, then

$$F_{\delta\pm}(\varphi,\Delta)f = 0 .$$

Proof. We refer to [2] Lemma 4.3.

Lemma 4.3. For all δ and Δ

$$F_{\delta\pm}(\varphi,\Delta) = F_{\delta\pm}(\varphi,\Delta)E_{\delta}(\varphi,\Delta)$$

If $\tilde{\Delta} \subseteq \Delta$, then for $f \in E_{\delta}(\varphi,\tilde{\Delta})H$

$$F_{\delta\pm}(\varphi,\Delta)f = F_{\delta\pm}(\varphi,\tilde{\Delta})f \ .$$

Proof. Consider $F_{\delta+}$. By Theorem 3.3, for $f \in H$

$$E_{\delta}(\varphi,\Delta)(f - E_{\delta}(\varphi,\Delta)f) = 0 \ ,$$

hence by Lemma 4.2

$$F_{\delta\pm}(\varphi,\Delta)(f - E_{\delta}(\varphi,\Delta)f) = 0 \ .$$

If $\tilde{\Delta} \subseteq \Delta$ and $f = E_{\delta}(\varphi,\tilde{\Delta})f$, then

$$F_{\delta\pm}(\varphi,\Delta\setminus\tilde{\Delta})f = F_{\delta\pm}(\varphi,\Delta\setminus\tilde{\Delta})E_{\delta}(\varphi,\Delta\setminus\tilde{\Delta})E_{\delta}(\varphi,\tilde{\Delta})f = 0$$

and hence

$$F_{\delta\pm}(\varphi,\Delta)f = F_{\delta\pm}(\varphi,\tilde{\Delta})f \ .$$

Lemma 4.4. For all δ and Δ , the operators $F_{\delta\pm}(\varphi,\Delta)$ map $E_{\delta}(\varphi,\Delta)H$ onto $L^2(\Delta,h_{\delta})$.

Proof. We refer to [2] Theorem 4.5.

We can now define the local wave operators $W_{\delta\pm}(\varphi,\Delta)$ for $0 < \varphi < a$ and all δ , Δ by

$$W_{\delta\pm}(\varphi,\Delta) = F_{\delta\pm}^{-1}(\varphi,\Delta)F_{\delta} \tag{4.5}$$

We set

$$W_{\delta\pm}(\varphi) = W_{\delta\pm}(\varphi,\mathbb{R}^+) = W_{\delta\pm}(\varphi,\mathbb{R}) \tag{4.6}$$

Theorem 4.5. For $\alpha \neq 0$, $W_{\pm}(\varphi,\Delta)$ maps $E_{\alpha,0}(\Delta)L^2(\mathbb{R}^3_{y_\alpha})$ one-to-one and bicontinuously onto $E_{\alpha}(\varphi,\Delta)H$. Moreover, for $\tilde{\Delta} \subseteq \Delta$

$$E_{\alpha}(\varphi,\tilde{\Delta})W_{\alpha\pm}(\varphi,\Delta) = W_{\alpha\pm}(\varphi,\Delta)E_{\alpha,0}(\tilde{\Delta}) \tag{4.7}$$

where $E_{\alpha,0}(\Delta)$ is the spectral measure of $-\frac{\Delta y_\alpha}{2n_\alpha}$.

The same holds for $\alpha = 0$ with $L^2(\mathbb{R}^3_{y_\alpha})$ replaced by H, $\{E_0(\Delta)\}$ being the spectral measure of H_0.

Proof. The first statement follows from Lemmas 4.1 and 4.4. To prove the intertwining relation we note that (4.7) is equivalent to

$$F_{\alpha\pm}(\varphi,\Delta)E_\alpha(\varphi,\tilde{\Delta}) = F_\alpha E_{\alpha,0}(\tilde{\Delta})F_\alpha^{-1} F_{\alpha\pm}(\varphi,\Delta) \tag{4.8}$$

By Lemma 4.3, the left hand side of (4.8) equals $F_{\alpha\pm}(\varphi,\tilde{\Delta})$. Moreover,

$$F_\alpha E_{\alpha,0}(\varphi,\tilde{\Delta})F_\alpha^{-1} = \chi_{\{\xi \mid |\xi|^2/n_\alpha \in \tilde{\Delta}\}},$$

hence the right hand side equals $F_{\alpha\pm}(\varphi,\tilde{\Delta})$, and the Theorem is proved.

5. Limits of spectral projections and wave operators as $\varphi \to 0_+$.

In this section we make the following uniformity assumptions on V_α for some $s > 1$, $p > \frac{3}{2}$

A 1 u) (i) For every $\varphi \in I$, $V(\varphi)$ has a decomposition

$$V(\varphi) = V_1(\varphi) + V_2(\varphi)$$

such that

$$V_1(\varphi) \in L^p \cap L^1_{2s},\ V_2(\varphi) \in L^\infty_{2s}.$$

The function $V_1(\varphi)$ is continuous from $(-a,a)$ into L^p and from $(-a,a)$ into L^1_{2s}, while $V_2(\varphi)$ is continuous from $(-a,a)$ into L^∞_{2s}.

Moreover, there exists for every $\varphi_0 \in (-a,a)$ $\delta(\varphi_0) > 0$ and $F_{\varphi_0} \in L^p \cap L^1_{2s}$ such that

$$|V_1(x)| \leq F_{\varphi_0}(x) \quad \text{for} \quad |\varphi-\varphi_0| \leq \delta(\varphi_0),\ x \in \mathbb{R}^3.$$

(ii) For $\alpha = (12),(23),(31)$ the operators h_α have no positive eigenvalues and for $0 < \varphi \leqq \varphi_0$ the operators $h_\alpha(\varphi)$ have no resonances.

A 2 u) (i) $V(\varphi)$ satisfies A 1 u) (i) with p replaced by $2p$ and and L^1_{2s} replaced by L^2_{2s} .

(ii) identical with A 1 u) (ii).

Remark. The absence of positive eigenvalues has been proved under weak regularity conditions on V . Moreover, assumption C implies that there exists $\varphi_0 > 0$ such that $h_\alpha(\varphi)$ have no resonances for $0 < \varphi \leqq \varphi_0$ (cf. Remark 1.5).

Under these conditions we shall now investigate the limit $\varphi \to 0_+$, utilizing the uniform estimates of [3], sections 4 and 5.

The following result is proved in [2] Lemmas 5.8 and 5.11:

Lemma 5.1. Suppose that A 1 u) is satisfied and that $H(\varphi)$ have no resonances in $\bigcup_{0<\varphi\leqq\varphi_0} \{\lambda_\alpha + e^{-2i\varphi}\,\mathbb{R}^+\}$ and that H has no eigenvalues or threshold resonances in $[\lambda_\alpha, \lambda_\alpha']$. Then for $f \in \mathcal{H}$ the following limits exists in $L^2((0,\lambda_\alpha' - \lambda_\alpha) ; L^2_s(\mathbb{R}^3_{y_\alpha}))$,

$$\lim_{\varphi \to 0_+} Y_{\alpha+}(\varphi , \lambda_\alpha + e^{-2i\varphi}\mu) f = Y'_{\alpha+}(\lambda_\alpha + \mu) f \; . \tag{5.1}$$

Suppose that A 2 u) is satisfied and that $H(\varphi)$ have no resonances in $\bigcup_{0<\varphi\leqq\varphi_0} \{e^{-2i\varphi}\,\overline{\mathbb{R}^+}\}$, no zero - energy resonance and no positive eigenvalues. Then the following limit exists in $L^2(\mathbb{R}^+ ; X)$ for $f \in \mathcal{H}$,

$$\lim_{\varphi \to 0_+} Y_{0+}(\varphi , e^{-2i\varphi}\mu) f = Y'_{0+}(\mu) f \; . \tag{5.2}$$

<u>Theorem 5.2</u>. Under the assumptions of Lemma 5.1 the following limits exist in $L^2((\lambda'_\alpha - \lambda_\alpha ; h_\alpha))$ for every $f \in H$,

$$\lim_{\varphi \to 0_+} F_{\alpha+}(\varphi,[\lambda'_\alpha - \lambda_\alpha])f = F_{\alpha+}([\lambda'_\alpha - \lambda_\alpha])f \tag{5.3}$$

where $F_{\alpha+}([\lambda'_\alpha - \lambda_\alpha])$ is the α-element of $F_+([\lambda'_\alpha - \lambda_\alpha])$.

The h_α-valued function

$$z = \rho^{-1}e^{i\varphi} \to g_\alpha(z) = T_\alpha(\rho^2)e^{-i\varphi/2}\, Y_{\alpha+}(\varphi,\lambda_\alpha + e^{-2i\varphi}\rho^2)f(e^{i\varphi})$$

is analytic in z for $0 < \varphi < \varphi_0$, $0 < \rho < \infty$ and continuous for $0 \leqq \varphi \leqq \varphi_0$, $0 \leqq \rho \leqq (\lambda'_\alpha - \lambda_\alpha)^{\frac{1}{2}}$. For $0 < \varphi < \varphi_0$, $\mu \in \mathbb{R}^+$

$$(F_{\alpha+}(\varphi, \mathbb{R}^+)f)(\mu) = g_\alpha(\mu^{-\frac{1}{2}}e^{i\varphi}) \tag{5.4}$$

For $\varphi = 0$, $\mu \in [0,\lambda'_\alpha - \lambda_\alpha]$,

$$(F_{\alpha+}([\lambda'_\alpha - \lambda_\alpha])f)(\mu) = g_\alpha(\mu^{-\frac{1}{2}}) \tag{5.5}$$

For $\alpha = 0$ the following limits exists in $L^2(\mathbb{R}^+, h_0)$ for every $f \in H$,

$$\lim_{\varphi \to 0_+} F_{0+}(\varphi,\mathbb{R}^+)f = F_{0+}(\mathbb{R}^+)f \tag{5.6}$$

where $F_{0+}(\mathbb{R}^+)$ is the zero-element of the local inverse wave operator $F_+(\mathbb{R}^+)$.

Let f be an S_a-dilation analytic vector in $L^2_s(\mathbb{R}^6)$. The h_0-valued function

$$z = \rho^{-1}e^{i\varphi} \to T_0(\rho^2)e^{-2i\varphi}\, Y_{0+}(\varphi,e^{-2i\varphi}\rho^2)f(e^{i\varphi}) = g_0(z)$$

is analytic in z for $0 < \varphi < \varphi_0$, $0 < \rho < \infty$ and continuous for $0 \leqq \varphi \leqq \varphi_0$, $0 \leqq \rho < \infty$; for $0 \leqq \varphi \leqq \varphi_0$

$$(F_{0+}(\varphi,\mathbb{R}^+)f(e^{i\varphi}))(\mu) = \chi_{\mathbb{R}^+}(\mu)\, g_0(\mu^{-\frac{1}{2}}e^{i\varphi}) . \tag{5.7}$$

<u>Proof</u>. We prove the case $\alpha = 0$, the case $\alpha \neq 0$ is similar.

It follows from Lemmas 1.6 and 5.1 that in $L^2(\mathbb{R}^+, h_0)$

$$\lim_{\varphi \to 0_+} \chi_{\mathbb{R}^+}(\mu)\, T_0(\mu)\, Y'_{0+}(\varphi,e^{-2i\varphi}\mu)f = \chi_{\mathbb{R}^+}(\mu)\, T_0(\mu)\, Y'_{0+}(\mu)f ,$$

and (5.6) follows.

Let $f \in L^2_s(\mathbb{R}^6)$ be S_a - dilation - analytic. Then, with

$c_0 = \frac{1}{2}(2m)^{3/2}$

$$\begin{aligned} g_0(z) &= c_0 \rho^2 e^{-2i\varphi} \gamma_0(\rho(2m)^{\frac{1}{2}}) F Y_{0+}(\varphi, e^{-2i\varphi}\rho^2) f(e^{i\varphi}) \\ &= c_0 z^{-2} \gamma_0((2m)^{\frac{1}{2}}) U(\rho) \quad F Y_{0+}(\varphi, e^{-2i\varphi}\rho^2) f(e^{i\varphi}) \\ &= c_0 z^{-2} \gamma_0((2m)^{\frac{1}{2}}) F U(\rho^{-1}) Y_{0+}(\varphi, e^{-2i\varphi}\rho^2) U(\rho) U(\rho^{-1}) f(e^{i\varphi}) \\ &= c_0 z^{-2} \gamma_0((2m)^{\frac{1}{2}}) F Y_{0+}(z, z^{-2}) f(z) \ . \end{aligned}$$

By [1] Lemma 4.18, $Y_{0+}(z,z^{-2})$ is a $\mathcal{B}(L^2_s(\mathbb{R}^6), X)$ - valued, analytic function of z, and $f(z)$ is by assumption an $L^2_s(\mathbb{R}^6)$ - valued analytic function of z in $\{z \mid 0 < \varphi < \varphi_0 \ , \ \rho > 0\}$, hence $g_0(z)$ is analytic in thie domain. By [1] Lemma 2.9, the following limit exists in the uniform operator topology for all $\mu \geqq 0$ such that μ is not an eigenvalue of H,

$$Y_{0+}(\mu) = \lim_{\varepsilon \downarrow 0} Y_0(\mu + i\varepsilon) \in \mathcal{B}(L^2_s(\mathbb{R}^6), X) \ .$$

It is easy to show that if ρ^2 is not an eigenvalue or zero - energy resonance of H, then

$$\lim_{\varphi \to 0_+} Y_{0+}(\varphi, e^{-2i\varphi}\rho^2) = Y_{0+}(\rho^2) \quad \text{in} \quad \mathcal{B}(L^2_s(\mathbb{R}^6), X) \ .$$

This implies the continuity of $g_0(z)$ for $0 \leqq \varphi \leqq \varphi_0$, $\rho \geqq 0$. The identity (5.7) holds by definition.

<u>Remark 5.3</u>. The weak limits of the operators $E_\alpha(\varphi, \Delta)$ as $\varphi \to 0_+$ exist; for $f, g \in \mathcal{H}$, $\Delta_\alpha = [\lambda_\alpha' - \lambda_\alpha]$ for $\alpha \neq 0$, $\Delta_0 = \overline{\mathbb{R}^+}$,

$$\lim_{\varphi \to 0_+} (f, E_\alpha(\varphi, \Delta_\alpha) g) = (\widetilde{F}_{\alpha+}(\Delta_\alpha) f \ , F_{\alpha+}(\Delta_\alpha) g)_{L^2(\Delta_\alpha, h_\alpha)}$$

where

$$\widetilde{F}_{\alpha+}(\Delta_\alpha) = \text{s} - \lim_{\varphi \to 0_+} F_{\alpha+}(-\varphi, \Delta_\alpha)$$

exists by similar arguments as used for the existence of

$\text{s} - \lim_{\varphi \to 0_+} F_{\alpha+}(\varphi, \Delta_\alpha)$.

The operator $\widetilde{F}_\alpha(\Delta_\alpha)$, however, is not related in any simple way to known operators; it involves both the limits $\widetilde{\gamma}_{\beta-}(\mu)$ (for $\lambda_\beta < \lambda_\alpha$) and $\widetilde{\gamma}_{\alpha+}(\mu)$ .The significance of the operators $\widetilde{F}_{\alpha+}(\Delta_\alpha)$ as well as the weak limits of $E_\alpha(\varphi,\Delta_\alpha)$, if any, is obscure. If they were proved to be strong limits, they would be projection operators with the same null space as $F_{\alpha+}(\Delta)$, but it is not clear whether they are in fact strong limits.

References.

1. E. Balslev, Analytic scattering theory of quantum-mechanical three-body Systems,
   a) Ann. Inst. H. Poincaré XXXII, 2 (1980), 125-160.
   b) Aarhus University, Preprint 1978/79 No.26.

2. E. Balslev, Wave operators for dilation-analytic three-body Hamiltonians,
   Aarhus University, Preprint 1986.

3. E. Balslev and E. Skibsted, Boundedness of two- and three-body resonances,
   Ann. Inst. H. Poincaré 43 , 4 (1985), 369-397.

4. J. Ginibre and M. Moulin, Hilbert space approach to the quantum-mechanical three-body problem,
   Ann. Inst. H. Poincaré XXI, 2 (1974), 97-145.

5. K. Hoffmann, Banach Spaces of Analytic Functions, London 1962.

6. T. Kato, Wave operators and similarity for some none-self-adjoint operators, Math. Ann. 162 (1966), 258 - 279.

7. C. van Winter, The resolvent of a dilation-analytic three-particle system,
   J. Math. Anal. Appl. 101 (1984), No.1, 195-267.

8. K. Yajima, An abstract stationary approach to three-body scattering,
   J. Fac. Sci. Univ. Tokyo, Sect. I A 25 (1978), 109-132.

INTRODUCTION TO ASYMPTOTIC OBSERVABLES FOR MULTIPARTICLE QUANTUM SCATTERING

Volker Enss

Institut für Mathematik I
Freie Universität Berlin
Arnimallee 2 - 6
D-1000 Berlin 33
West-Germany

Abstract. We give an expository presentation of the convergence proof for asymptotic observables in N-body quantum systems. Some applications are derived. We use simplifying assumptions and add numerous remarks to stress the main ideas and techniques and to avoid technicalities. Auxiliary concepts like "k-clustered operators" and local decay of subsystems are discussed in detail.

I Introduction.

In these notes we give the main parts of a convergence proof for suitable observables on scattering states for quantum mechanical N-body systems. Here we restrict ourselves to a simpler special case. This should help to stress the main concepts and ideas of the approach. Also the burden of many technicalities can be reduced. The complete proof under much weaker assumptions will be given in [4].

According to the experience abstracted from observations in physics an N-particle system should for late times behave like a state of several bounded clusters (e.g. atoms) moving freely relative to each other (or superpositions thereof in quantum mechanics). The free relative motion of subsystems has the property that at a late time τ the position $y(\tau)$ divided by τ, i.e. the average velocity $y(\tau)/\tau$, approaches the instantaneous velocity $q(\tau)$. The asymptotic correlations between positions and velocities (momenta) characterize a small subset of classical phase space which is "absorbing" for a scattering system.

For a decomposition d_k into k bounded subsystems let

$\tilde{P}(d_k) \cdot \exp(-iH\tau)\Psi$ denote the part of the state Ψ at time τ where just these subsystems are bounded. Then *roughly*

$$\left\{\frac{Y(d_k)}{\tau} - Q(d_k)\right\} \tilde{P}(d_k)\ e^{-iH\tau}\ \Psi \xrightarrow[\tau\to\infty]{} 0 \tag{1.1}$$

where $Y(d_k) \in \mathbb{R}^{\nu(k-1)}$ denotes the relative position operator of the subsystems and $Q(d_k)$ the corresponding velocity operator. A precise statement of this result and of its implications for the asymptotic phase space localization are given below as Theorem 3.1 and Corollary 3.2 after we have specified the notation. Other results of independent interest are Theorem 5.6 and Corollary 5.8.

The magic aspect of these results and their proofs is that they are elementary, although not simple. The main effects are purely kinematical and the forces between clusters of particles appear only as perturbations which vanish asymptotically. The reason for this is that we contend ourselves with convergence to zero in (1.1). Actually for the motion under short-range forces one expects a $1/\tau$ - decay and $(\ln \tau)/\tau$ for Coulomb forces. But these decay rates have to be proved by more subtle methods. A significant simplification of the proof comes from the possibility to obtain simple expressions by summing over terms which are individually unpleasant. This avoids e.g. to control the time evolution of $\tilde{P}(d_k)$ which would be quite complicated.

Assumptions. To simplify the notation we assume that the forces are described by local two-body potentials for each pair α of particles. The functions $V^\alpha(x^\alpha)$ are assumed to be bounded C^1-functions with

$$V^\alpha(x^\alpha) \to 0 \qquad \text{as} \quad |x^\alpha| \to \infty\ . \tag{1.2}$$

For their derivatives we require

$$x^\alpha \cdot (\nabla V^\alpha)(x^\alpha) \to 0 \qquad \text{as} \quad |x^\alpha| \to \infty\ . \tag{1.3}$$

The inclusion of singular short-range potentials is a purely technical exercise and so we omit it here.

The other assumption for convenience is more restrictive. We suppose that the system and all its subsystems have *finitely many bound states*. This shortens the notation and simplifies the statement of our results. It turns out that in the general case one has to choose inductively a finite number of bound states for the subsystems depending on the decomposition d_k and the admissible error ε. Technically this is better than the conventional splitting into the continuous and point-spectral subspaces. For details see [4].

Moreover we assume that all eigenfunctions have suitable power-decay. This has to be assumed only if there happen to be eigenvalues at thresholds. Otherwise exponential decay is known [5]. In the general case one avoids this assumption by using a spatial cutoff for the eigenfunctions [4].

Asymptotic observables are a useful tool in proofs of asymptotic completeness for two-body Schrödinger and Dirac operators. They are crucial for the geometrical time-dependent approach to three-body systems, see e.g. [1 - 4, 6, 9, 10].

Acknowledgements. These are the considerably expanded notes of a lecture given at the Symposium. I am very grateful to Erik Balslev for organizing this very fruitful meeting and to the participants for vivid discussions.

II. Notation.

We introduce the notation to describe N-body systems. We recommend that the reader proceeds directly to the next section and refers back to the definitions when the need arises.

Decompositions d_k of the set of particles into k non-empty, pairwise disjoint clusters c_j (subsystems):

$$d_k = \{c_1, c_2, \ldots, c_k\} \ , \quad \bigcup_j c_j = \{1, 2, \ldots, N\} \ . \tag{2.1}$$

Summation over d_k means summation over all decompositions with *fixed* k. $|c_j|$ is the number of particles in c_j.

Coordinates and inner product. It is convenient to use clustered Jacobi coordinates (see e.g. [8]). The particles move in ν-dimensional space. For a given cluster c_j with $|c_j| \geq 2$ assign somehow numbers r to the particles, $1 \leq r \leq |c_j|$; x_r is the position of the r-th particle. The relative position of all particles in the cluster is given by a $\nu(|c_j| - 1)$-dimensional vector $X(c_j)$. E.g. its first ν

components are $x_2 - x_1$, the next ones are the position of the third particle relative to the center of mass of the pair, i.e.

$$x_3 - (m_1 x_1 + m_2 x_2)/(m_1 + m_2) \ ,$$

and so on. To this order corresponds a $\nu(|c_j| - 1)$-dimensional diagonal mass matrix $M(c_j)$ with the reduced mass of the first pair $m_1 \cdot m_2/(m_1 + m_2)$ as the first ν entries, $m_3 \cdot (m_1 + m_2)/(m_1 + m_2 + m_3)$ as the next ν etc.

A convenient inner product is

$$\begin{aligned} &< X'(c_j), X(c_j) > := X'(c_j) \cdot M(c_j)\ X(c_j) \ , \\ &|X(c_j)|^2 := < X(c_j), X(c_j) > \ , \end{aligned} \tag{2.2}$$

where $\cdot$ denotes the Euclidean scalar product. While $X(c_j)$ and $M(c_j)$ depend also on the chosen order, the inner product (2.2) depends only on the relative configuration of the particles in c_j. We set $X(c_j) \equiv 0$ if $|c_j| = 1$.

$\mathcal{H}(c_j) \cong L^2(\mathbb{R}^{\nu(|c_j|-1)}, dX(c_j))$ is the <u>internal state space</u> of c_j.

Similarly the $\nu(k-1)$-dimensional vector $Y(d_k)$ describes the relative position of the centers of mass of the clusters where the role of the particle masses is played by the total masses of the clusters. $M(d_k)$ is defined analogously and

$$\begin{aligned} &< Y'(d_k), Y(d_k) > := Y'(d_k) \cdot M(d_k)\ Y(d_k) \ , \\ &|Y(d_k)|^2 := < Y(d_k), Y(d_k) > \end{aligned} \tag{2.3}$$

depends only on the configuration. We use the same letters for the multiplication operators in configuration space.

Since we will separate off the trivial free total center of mass motion the collection $\{Y(d_k); X(c_j), c_j \in d_k\}$ or $Y := Y(d_N)$ determine the configuration completely. In these coordinates for any d_k

$$|Y|^2 \equiv |Y(d_N)|^2 = |Y(d_k)|^2 + \sum_{c_j \in d_k} |X(c_j)|^2 \ . \tag{2.4}$$

The <u>Hilbert space</u> is (isomorphic for any d_k):

$$H \cong L^2(\mathbb{R}^{\nu(N-1)}, dY) \cong L^2(\mathbb{R}^{\nu(k-1)}, dY(d_k)) \bigotimes_{c_j \in d_k} H(c_j) .$$

<u>Velocities and momenta</u>. The canonically conjugate momentum to the intercluster-position $Y(d_k)$ is

$$-i \ \mathrm{grad}_{Y(d_k)} . \tag{2.5}$$

The corresponding velocity operator is

$$Q(d_k) := M(d_k)^{-1} (-i) \ \mathrm{grad}_{Y(d_k)} . \tag{2.6}$$

With the kinetic energy operators H_0 and $T(d_k)$ as given below it satisfies the identities

$$Q(d_k) = i[H_0, Y(d_k)] = i[T(d_k), Y(d_k)] . \tag{2.7}$$

<u>Hamiltonians</u>. All Hamiltonians are self-adjoint on natural domains. The internal *free cluster Hamiltonians* are

$$H_0(c_j) := -\frac{1}{2} \ \mathrm{grad}_{X(c_j)} \cdot M(c_j)^{-1} \ \mathrm{grad}_{X(c_j)} \tag{2.8}$$

i.e. up to mass factors $\nu(|c_j|-1)$-dimensional Laplacian operators. $H_0(c_j) = 0$ if $|c_j| = 1$. The kinetic energy of the intercluster motion is

$$\begin{aligned} T(d_k) &:= \frac{1}{2} \langle Q(d_k), Q(d_k) \rangle \\ &= -\frac{1}{2} \ \mathrm{grad}_{Y(d_k)} \cdot M(d_k)^{-1} \ \mathrm{grad}_{Y(d_k)} . \end{aligned} \tag{2.9}$$

The *free Hamiltonian* is for any d_k

$$H_0 := T(d_N) = T(d_k) + \sum_{c_j \in d_k} H_0(c_j) . \tag{2.10}$$

All these operators do not depend on the particular clustered Jacobi coordinates (order of the particles).

The interactions are given by multiplication with pair potentials V^α where α labels the pairs. Here they are assumed to satisfy (1.2) and for some estimates also (1.3). $\alpha \in d_k$ means $\alpha \subset c_j \in d_k$ for some j . The interacting *cluster Hamiltonians* are

$$H(c_j) = H_0(c_j) + \sum_{\alpha\subset c_j} V^\alpha \tag{2.11}$$

and the *decomposition Hamiltonian* is

$$H(d_k) = T(d_k) + \sum_{c_j\in d_k} H(c_j) = H_0 + \sum_{\alpha\in d_k} V^\alpha . \tag{2.12}$$

The fully *interacting Hamiltonian* is denoted by H

$$H = H_0 + \sum_{\text{all }\alpha} V^\alpha = H(d_1) \tag{2.13}$$

$$= H(d_k) + \sum_{\alpha\notin d_k} V^\alpha .$$

It generates the time evolution $\exp(-iHt)$ which has to be analysed.

We do not distinguish in our notation between operators acting on a factorspace like $\mathcal{H}(c_j)$ and their natural action on $\mathcal{H}$.

Bound state projections. We assumed above that each $H(c_j)$ has only finitely many eigenvectors (in its subsystem Hilbert space $\mathcal{H}(c_j)$). The corresponding projection (on the whole space $\mathcal{H}$) is

$$P(c_j) := \begin{cases} P^{pp}(H(c_j)) & \text{if } |c_j| \geq 2 , \\ \mathbb{1} & \text{if } |c_j| = 1 . \end{cases} \tag{2.14}$$

To specify that all subsystems are bounded we use

$$P(d_k) := \prod_{c_j\in d_k} P(c_j) . \tag{2.15}$$

The decay assumption for bound states which we use for simplicity is

$$\| |X(c_j)|^2 P(c_j)\| < \infty \quad \text{for all clusters } c_j . \tag{2.16}$$

$$\hat{P}(k) := [\mathbb{1} - \sum_{d_k} P(d_k)] \ \ldots \ [\mathbb{1} - \sum_{d_2} P(d_2)][\mathbb{1} - P(d_1)] \tag{2.17}$$

excludes all arrangements with k or less bounded subsystems. Clearly $\hat{P}(n) = 0$, $\hat{P}(1) = P^{cont}(H)$, and we set $\hat{P}(0) := \mathbb{1}$.

We introduce the following partition of the identity

$$\widetilde{P}(d_k) = P(d_k)\ \hat{P}(k-1) \ , \quad \sum_k \sum_{d_k} \widetilde{P}(d_k) = \mathbb{1} \ , \tag{2.18}$$

which specifies the states where precisely the clusters of d_k are bounded but no bigger subsystems. (The deviation from pairwise orthogonal projections decays in long time averages and thus does not matter below.)

Dilation operators. The group of dilations (scale transformations) of the intercluster coordinates $Y(d_k)$ is generated by

$$D(d_k) = \frac{1}{2}\{ < Q(d_k), Y(d_k) > + < Y(d_k), Q(d_k) > \} \ . \tag{2.19}$$

In particular $D \equiv D(d_N)$ is the usual dilation operator for all coordinates. Denoting the corresponding operator for the clusters by $D(c_j)$ we have for any d_k

$$D = D(d_k) + \sum_{c_j \in d_k} D(c_j) \ . \tag{2.20}$$

$D(c_j)$ is defined and essentially self-adjoint on $\mathcal{D}(|X(c_j)|^2) \cap \mathcal{D}(H(c_j))$ and $\|D(c_j)P(c_j)\| < \infty$ follows from (2.16). We get as quadratic forms

$$i[H_0(c_j), D(c_j)] = 2\ H_0(c_j) \ . \tag{2.21}$$

Analogously for any d_k

$$i[T(d_k), D(d_k)] = 2\ T(d_k) \ . \tag{2.22}$$

The sets of "cluster relatively compact" operators $RC(\cdot)$ are defined at the beginning of Section IV.

III. The Main Result and its Implications.

Now we state precisely the convergence indicated in (1.1).

Theorem 3.1. For any $\Psi \in \mathcal{H}$ there is a sequence of times $\tau_n \to \infty$ (and $\to -\infty$) such that for all $g \in C_0^\infty(\mathbb{R})$, $f_k \in C_0^\infty(\mathbb{R}^{\nu(k-1)})$ and all k, d_k

$$\| [g(H) - g(H(d_k))]\ \widetilde{P}(d_k)\ e^{-iH\tau_n}\ \Psi \| \to 0 \ , \tag{3.1}$$

$$\|\{f_k\left(\frac{Y(d_k)}{\tau_n}\right) - f_k(Q(d_k))\}\ \tilde{P}(d_k)\ e^{-iH\tau_n}\ \psi\ \| \to 0 \tag{3.2}$$

as $n \to \infty$.

Here $\tilde{P}(d_k)$ specifies that all clusters in the decomposition d_k are bounded but not any bigger subsystem (for the precise definition see *bound state projections* (2.18) in Section II). $Y(d_k)$ denotes the relative positions of the centers of mass of the clusters and $Q(d_k)$ their velocities (c.f. *coordinates*, *velocities*, resp.). H is the fully interacting Hamiltonian (2.13) and $H(d_k)$ the decomposition Hamiltonian (2.12) with intercluster-interactions omitted (see *Hamiltonians*). The statement is non-vacuous for $k \geq 2$.

On physical grounds one expects that (3.1) and (3.2) hold in the limit $|\tau| \to \infty$ and not only for subsequences. It would follow e.g. from asymptotic completeness. For $N = 2$ it follows with our approach [2], but it seems that for $N \geq 3$ more subtle estimates would have to be used. In our proof of asymptoic completeness for three particle systems [3] the existence of a sequence τ_n was sufficient and we hope that the same applies to higher N .

(3.1) is essentially equivalent to the fact that the clusters in d_k seperate for late times and therefore the intercluster potentials do not affect the motion asymptotically. For the simpler separation alone see e.g. Corollary 5.8 below.

The support of f_k may be arbitrarily small. For the applications one may think of f_k as a smoothed characteristic function. We deduce from (3.2) that at a late time τ_n
$Y(d_k) - \tau_n\, Q(d_k) \lesssim (\text{diam supp } f_k)\, \tau_n$. Due to the increase of the size of the localization region in space there is no conflict with the uncertainty principle. For f_k with $f_k(0) = 1$ property (3.2) can equivalently be replaced by

$$\|\{f_k\left(\frac{Y(d_k)}{\tau_n} - Q(d_k)\right) - \mathbb{1}\}\ \tilde{P}(d_k)\ e^{-iH\tau_n}\ \psi\ \| \tag{3.3}$$

$$= \|\{e^{-iH(d_k)\tau_n}\ f_k\left(\frac{Y(d_k)}{\tau_n}\right)\ e^{+iH(d_k)\tau_n} - \mathbb{1}\}\ \tilde{P}(d_k)\ e^{-iH\tau_n}\ \psi\| \to 0\ . \tag{3.4}$$

Thus the phase space localization of the state at τ_n is such as if the clusters had moved freely relative to each other after starting at time zero inside a ball of radius (diam supp f_k) τ_n. One cannot expect any better phase space localization because we admitted potentials of very slow decay. There are classical trajectories which essentially saturate these bounds. The same should hold for quantum states.

In scattering theory one is interested in approximations of the full time evolution by simpler ones for late times. In multichannel scattering one has to decompose a state at late times into components where the different asymptotic time evolutions should be good approximations. The above result may be useful to give such a splitting. Consider e.g. the dense set of states Ψ with $\Psi = g(H)\Psi$ where $g \in C_0^\infty(\mathbb{R})$ and the support of g does not contain thresholds. Then for late times by (3.1)

$$e^{-iH\tau}\Psi = g(H) \sum_k \sum_{d_k} \tilde{P}(d_k)\, e^{-iH\tau}\,\Psi$$

$$\approx \sum_k \sum_{d_k} g(H(d_k))\, \tilde{P}(d_k)\, e^{-iH\tau}\,\Psi$$

$$\approx \sum_m \sum_k \sum_{d_k} [f_k^{(m)}(Q(d_k))]^2\, \tilde{P}(d_k)\, e^{-iH\tau}\,\Psi \;.$$

The last relation holds whenever

$$\sum_m [f_k^{(m)}]^2 = 1$$

on the compact subset of $Q(d_k)$-space in the range of $g(H(d_k))\, P(d_k)$. m should run over a finite index set. With (3.2) we get

<u>Corollary 3.2.</u> Under the above assumptions

$$\lim_{n\to\infty} \| e^{-iH\tau_n}\Psi - \sum_m \sum_k \sum_{d_k} f_k^{(m)}\left(\frac{Y(d_k)}{\tau_n}\right) f_k^{(m)}(Q(d_k))\, \tilde{P}(d_k)\, e^{-iH\tau_n}\Psi \| = 0 \;.$$

Thus the old scattering state is decomposed into finitely many pieces. On the range of

$$f_k^{(m)}\left(\frac{Y(d_k)}{\tau}\right) f_k^{(m)}(Q(d_k))\, P(d_k)$$

the clusters are bounded and the state is localized in the classical phase space for the relative motion of the clusters. For $N = 3$

particles and a suitable choice of the functions $f_k^{(m)}$ this led to a proof of asymptotic completeness for potentials of short and of long range [3].

IV. Clustered Operators.

A bounded operator K is called <u>d_k-clustered</u> or <u>d_k-relatively compact</u>, denoted $K \in RC(d_k)$, if

$$\| K\, F(|X(c_j)| > R) \| \to 0 \quad \text{as} \quad R \to \infty \quad \text{for } \textit{all} \quad c_j \in d_k \ . \tag{4.1}$$

$X(c_j)$ are internal coordinates of the cluster c_j , see Section II. The set $RC(d_k)$ does not depend on their particular choice. No condition is imposed for clusters with $|c_j| = 1$ because then $X(c_j) \equiv 0$. Prototypes of such operators are $\prod_{c_j \in d_k} F(|X(c_j)| \le \rho)$, any $\rho < \infty$, or the bound state projections $P(d_k)$, (2.15). A parametrized family K_u is "uniformly in $RC(d_k)$" if for any $\varepsilon > 0$ there is an $R(\varepsilon)$ independent of u such that the norm in (4.1) is bounded by ε for all u .

Let $K_n \to K$ be a norm-convergent sequence with $K_n \in RC(d_k)$, then also $K \in RC(d_k)$. If $B(t)$ is a uniformly bounded family of operators, then $B(t)\,K \in RC(d_k)$ uniformly in t .

If for given k and for any $\varepsilon > 0$ there is a decomposition

$$\| K - \sum_{d_k} K(d_k) \| < \varepsilon \ , \quad K(d_k) \in RC(d_k) \tag{4.2}$$

then K is called <u>k-clustered</u> or <u>k-relatively compact</u>, denoted as $K \in RC(k)$. It is easy to see that $RC(k) \supset RC(k-1)$, $RC(N)$ is the set of all bounded operators. For uniform families analogously.

For $k = 1$ the condition $K \in RC(d_1) \equiv RC(1)$ implies compactness of $K(H-z)^{-1}$. [z will always be chosen in the resolvent set for all Hamiltonians.] Thus for $k > 1$ our notion extends a concept related to relative compactness for bounded operators. It can easily be generalized to include certain unbounded operators. Related sets have been used in [7].

Bounded decaying functions of the energy do not spoil the localization. This suggests

Lemma 4.1. Let f be a bounded continuous real or complex valued function on the real line with $\lim f(\omega) = 0$ as $|\omega| \to \infty$. Let H' be any Hamiltonian $H(d_1)$ etc., then $K \in RC(d_k)$ implies $K\, f(H') \in RC(d_k)$, and $K \in RC(k)$ implies $K\, f(H') \in RC(k)$.

Proof. Let $K \in RC(d_k)$. For every $c_j \in d_k$ we study

$$\|K\, f(H')\, F(|X(c_j)| > R)\| \leq \|K\, F(|X(c_j)| > R/2)\| \cdot \|f(H')\| +$$
$$+ \|K\| \cdot \|F(|X(c_j)| < R/2)\, f(H')\, F(|X(c_j)| > R)\| \quad . \tag{4.3}$$

It is sufficient to estimate the second summand. For

$$g_R(X(c_j)) = \begin{cases} 1 & \text{if } |X(c_j)| < R/2 \\ 0 & \text{if } |X(c_j)| > R \end{cases} \tag{4.4}$$

and $\sup |\nabla g_R| \leq \mathrm{const}/R$ we obtain the bound

$$\|F(\ldots < R/2)\,[g_R(X(c_j)), (H'-z)^{-1}]\; F(\ldots > R)\|$$

for f replaced by a resolvent. This is bounded by

$$\|[g_R(X(c_j)), (H'-z)^{-1}]\|$$
$$= \|(H'-z)^{-1}[g_R(X(c_j)), H_0](H'-z)^{-1}\| \leq \mathrm{const}/R \; . \tag{4.5}$$

Here we have used that the potentials commute with g_R and that linear functions of the momenta are bounded by resolvents. By iteration we get decay for any finite power of the resolvent and also for z replaced by $\bar{z}$. By the Stone-Weierstrass theorem the function f can be approximated arbitrarily well by a finite linear combination of powers of these resolvents. For $K \in RC(k)$ we use first (4.2). □

Corollary 4.2. With f, H' as above, $K \in RC(d_k)$

$$K\, e^{-iH's}\, f(H') \in RC(d_k) \quad \text{uniformly in } |s| \leq S < \infty \; .$$

Analogously for $RC(k)$.

Proof. $\exp(-iH's)\, f(H')$ is norm continuous. Thus the operators can be well approximated by some

$$K\, e^{-iH's_j}\, f(H') \tag{4.6}$$

where $|s_j| \leq S$ is chosen from a finite collection of times. With the above lemma for f times the exponential function the operator (4.6) lies in $RC(d_k)$ for any j. It is uniform for the finitely many. □

The following lemmas show that certain differences of operators are better connected (i.e. less clusters) than the individual terms.

Lemma 4.3. With f as above, $K \in RC(d_k)$

(a) $$[f(H) - f(H(d_k))]\, P(d_k) \in RC(k-1) \,, \tag{4.7}$$

(b) $$K\,[f(H) - f(H(d_k))] \in RC(k-1) \,. \tag{4.8}$$

Proof. (a) For resolvents

$$[(H-z)^{-1} - (H(d_k)-z)^{-1}]\, P(d_k) = -(H-z)^{-1} \sum_{\alpha \notin d_k} V^\alpha\, P(d_k)\, (H(d_k)-z)^{-1} \,.$$

For each $\alpha \notin d_k$ we have $V^\alpha\, P(d_k) \in RC(d_k \cup \alpha) \subset RC(k-1)$. To see this (intuitively obvious) fact it is sufficient to show decay if $|y^\alpha| \to \infty$, where y^α is the relative coordinate of the centers of mass of those clusters $c_j, c_j' \in d_k$ which are linked by α.

$$\begin{aligned} &\| V_\alpha\, P(d_k)\, F(|y^\alpha| > 3R) \| \\ &\leq \| V_\alpha\, F(|y^\alpha| > 3R)\, F(|X(c_j)| < mR)\, F(|X(c_j')| < mR) \| \\ &+ \| V_\alpha \| \, \{ \| F(|X(c_j)| > mR)\, P(d_k) \| + \| F(|X(c_j')| > mR)\, P(d_k) \| \} \,. \end{aligned} \tag{4.9}$$

On the range of the projections in the first term $|X(\alpha)| \equiv |x^\alpha| > R$ if $m > 0$ is chosen small enough depending on the masses which are hidden in the modulus $|\cdot|$, see (2.2), (2.3). The decay follows from (1.2). The remaining terms decay by the localization of bound states. A decay rate like (2.16) is not needed here. Similarly for higher powers of the resolvents and for their adjoints. With the Stone-Weierstrass theorem part (a) of the lemma follows.

$$\begin{aligned} &K\,[(H-z)^{-1} - (H(d_k)-z)^{-1}] \\ &= K\,(H(d_k)-z)^{-1} \sum_{\alpha \notin d_k} V^\alpha\, (H-z)^{-1} = \sum_{\alpha \notin d_k} K'\, V^\alpha\, (H-z)^{-1} \,. \end{aligned}$$

By Lemma 4.1 $K' \in RC(d_k)$ and thus $K'\, V^\alpha \in RC(d_k \cup \alpha) \subset RC(k-1)$. As above we conclude part (b). □

Corollary 4.4. With f as above and $\hat{P}(\cdot)$ defined in (2.17)

$$[f(H)\, e^{-iHs}, \hat{P}(k)] = \sum_{l=1}^{k-1} K_l(s)\, \hat{P}(l) \tag{4.10}$$

where $K_l(s) \in RC(l)$ uniformly in $|s| \le S < \infty$.

Proof. Again it is sufficient to study a finite collection of times s or a fixed time s . The function $f'(H) = f(H)\exp(-iHs)$ has the same properties as f .

$$[f'(H),\hat{P}(k)] = [f'(H),\{\mathbb{1} - \sum_{d_k} P(d_k)\}]\, \hat{P}(k-1)$$

$$+ \{\mathbb{1} - \sum_{d_k} P(d_k)\}\, [f'(H),\hat{P}(k-1)] \ .$$

$$[f'(H),\{\ldots\}] = -\sum_{d_k} [f'(H),P(d_k)] \ .$$

$$[f'(H),P(d_k)] = [f'(H)-f'(H(d_k))]\, P(d_k) - P(d_k)\, [f'(H)-f'(H(d_k))] + 0$$

$$\in RC(k-1)$$

by Lemma 4.3. Then (4.10) follows by iteration. □

Corollary 4.5. Let $K \in RC(d_k)$ and f as above, then

$$K\,\{\mathbb{1} - e^{-iH(d_k)s}\,e^{iHs}\}\,f(H) \in RC(k-1) \tag{4.11}$$

uniformly in $|s| \le S < \infty$.

Proof. The quantity to be studied is approximated in norm by

$$K\,\{e^{-iHs}\,g_E(H) - e^{-iH(d_k)s}\,g_E(H(d_k))\}\,f(H)\,e^{iHs}$$

where g_E is a family of uniformly bounded continuous functions decaying towards infinity with $g_E(\omega) = 1$ for $|\omega| \le E$. Then

$$\lim_{E\to\infty} \{\|[\mathbb{1} - g_E(H)]f(H)\| + \|[\mathbb{1} - g_E(H(d_k))]f(H)\|\} = 0$$

and the claim follows. For any finite E we have $K\{\ \} \in RC(k-1)$ by Lemma 4.3 b). The final factor $f(H)\cdot\exp(+iHs)$ does not spoil this by Corollary 4.2. □

Remark 4.6. There is a simple physical interpretation of the last corollary which makes clear that the restriction to finite time intervals and the high energy cutoff $f(H)$ are essential. The property of the operator to be in $RC(k-1)$ expresses the fact that either two clusters in d_k are not far separated or the operator is small. As long as all clusters in d_k are far separated the time evolutions generated by H or by $H(d_k)$ do not differ and the operator is small. This will fail if two clusters approach each other. If all clusters are separated at a given initial time then without an energy restriction they could get close in extremely short time. With an energy cutoff they may get close at some late time. But for any given error bound, finite time interval and high energy cutoff one can find a finite initial separation which is so large, that the clusters remain sufficiently far from each other for the given times because their maximal speed is bounded. One can justify this picture by carrying out the estimates correspondingly. However, the above procedure is shorter because it combines physically different arguments into a single estimate. Most of the other statements are variations of this theme.

So far we have provided the tools needed in the next section to show local decay. Up to now we have used only the simple decay (1.2). We conclude this section by showing cluster properties of some other terms arising in later sections in the study of local observables. We begin with a more general fact. We could have used this for another proof of Lemma 4.3 (a). We denote by $d_l \cup d_k$ the finest decomposition where any pair α is in a cluster if it is in a cluster for d_l or for d_k.

Lemma 4.7. Let $K \in RC(d_l)$ then $K\,P(d_k) \in RC(d_l \cup d_k)$.

Proof. Consider pairs α belonging to different clusters $c_j, c_j' \in d_k$ but to a single cluster of d_l. The claim is true if we show for each such α

$$\|K\ P(d_k)\ F(|y^\alpha| > R)\| \to 0 \quad \text{as} \quad R \to \infty\ .$$

By assumption $\|K\ F(|x^\alpha| > mR)\| \to 0$ and we have to estimate for suitably chosen $m > 0$

$$\|K\|\ \|F(|x^\alpha| < mR)\ F(|y^\alpha| > R)\ P(d_k)\|\ .$$

For small enough $m > 0$ depending on the masses the range of the projections is contained in the range of

$$F(|X(c_j)| > m'R) + F(|X(c_j')| > m'R)$$

for some $m' > 0$. The decay follows from the properties of $P(c_j)\cdot P(c_j')$. □

Corollary 4.8.

(a) $[\mathbb{1} - \hat{P}(k-1)^*]\ P(d_k) \in RC(k-1)\ ,\quad k \geq 2\ ,$

(b) $x^\alpha\cdot(\nabla V^\alpha)(x^\alpha)\ P(d_k) \in RC(k-1) \quad \text{for } \alpha \notin d_k\ .$

Proof. (a) $[\mathbb{1} - \hat{P}(k-1)^*]$ is a sum of non-empty products of bound state projections $P(d_l)$ with $l \leq k-1$ clusters.

(b) Identifying α with the decomposition d_{N-1} where the pair α is coupled and all other particles are singles we get $\alpha \cup d_k = d_{k-1}$. By (1.3) the bounded function $|x^\alpha\cdot(\nabla V^\alpha)(x^\alpha)|$ decays as $|x^\alpha| \to \infty$. □

Remark. This and the next lemmas are the only places where (1.3) is used. They will be applied in (7.4) in the proof of Theorem 3.1 and nowhere else.

Lemma 4.9. For any $c_j \in d_k$, $H(c_j)$ defined in (2.11), $D(c_j)$ in (2.20)

(a) $i[H(c_j), D(c_j)]\, P(d_k) \in RC(d_k)$,

(b) $f(H)\{e^{-iH(d_k)s} - e^{-iHs}\}\, i[H(c_j), D(c_j)]\, P(d_k) \in RC(k-1)$

uniformly in $|s| \le S < \infty$.

Proof. We calculate using (2.21) the commutator which is naturally defined as a quadratic form on a dense set:

$$\begin{aligned} i[H(c_j), D]\, P(c_j) &\equiv i[H(c_j), D(c_j)]\, P(c_j) \\ &= i[H_0(c_j), D(c_j)]\, P(c_j) + \sum_{\alpha \subset c_j} i[V^\alpha, D(c_j)]\, P(c_j) \\ &= 2H_0(c_j)\, P(c_j) - \sum_{\alpha \subset c_j} x^\alpha \cdot (\nabla V^\alpha)(x^\alpha)\, P(c_j) \\ &= \{2H(c_j) - \sum_{\alpha \subset c_j} [2V^\alpha + x^\alpha \cdot (\nabla V^\alpha)(x^\alpha)]\}\, P(c_j) . \end{aligned} \tag{4.12}$$

It extends to a bounded operator by the assumptions about the potentials. This implies statement (a). Moreover

$$\| F(|X(c_j)| > R)\, i[H(c_j), D(c_j)]\, P(d_k) \| \to 0 \quad \text{as} \quad R \to \infty \tag{4.13}$$

and the same property for the other clusters is trivial. For continuous decaying functions f also

$$[f(H) - f(H(d_k))]\, i[H(c_j), D(c_j)]\, P(d_k) \in RC(k-1) \tag{4.14}$$

as in the proof of Lemma 4.3 (a). Then (b) follows as in the proof of Corollary 4.5. □

Lemma 4.10. With $T(d_k)$ defined in (2.9)

$$(H-z)^{-1}\, [\mathbb{1} - \hat{P}(k-1)^*]\, T(d_k)\, P(d_k) \in RC(k-1) . \tag{4.15}$$

Proof. Specifying bound states in each cluster we get

$$P(d_k) = \sum_m P_m , \qquad \sum_{c_j \in d_k} H(c_j)\, P_m = E_m\, P_m ,$$

where the sum is finite. By Corollary 4.8 (a)

$$(H-z)^{-1}\, [\mathbb{1} - \hat{P}(k-1)^*] \sum_m E_m\, P_m \in RC(k-1) .$$

Therefore instead of (4.15) we can study equivalently

$$(H-z)^{-1} \; [\mathbb{1} - \hat{P}(k-1)^*] \; H(d_k) \; P(d_k) \; ,$$

or with $V^\alpha \; P(d_k) \in RC(k-1)$ for $\alpha \notin d_k$

$$(H-z)^{-1} \; [\mathbb{1} - \hat{P}(k-1)^*] \; H \; P(d_k) \; . \tag{4.16}$$

The left two factors are a sum of terms of the form

$$(H-z)^{-1} \; \{\prod P(d_r)\} \; P(d_l) \; , \quad l \leq k-1$$

where the product $\{\ldots\}$ may be empty, i.e. $\mathbb{1}$. Then $P(d_l)[H - H(d_l)] \; P(d_k) \in RC(k-1)$. With boundedness of $(H-z)^{-1} \; \{\prod P(d_r)\} \; (H(d_l)-z)$ the proof is complete. □

V. Improved Local Decay.

We first give a uniform variant of an estimate which is common in ergodic theory in its strong form. We adapt its form to the generalizations we have in mind. Recall that $\hat{P}(1) = P^{cont}(H)$ and for $K \in RC(1)$ the operator $K \; f(H)$ is compact for any bounded function f decaying towards infinity.

Proposition 5.1. Let $B(t)$ be a family of uniformly bounded operators: $\|B(t)\| \leq M < \infty$ for all $t \in \mathbb{R}$, $K \in RC(1)$. If the bounded real or complex valued function f satisfies $f(\omega) \to 0$ as $|\omega| \to \infty$, then

$$\lim_{|T|\to\infty} \left\| \frac{1}{T} \int_0^T dt \; B(t) \; K \; \hat{P}(1) \; e^{-iHt} \; f(H) \right\| = 0 \; . \tag{5.1}$$

The same holds uniformly for uniform families.

Remark 5.2. We have to assume that the dependence of $B(t)$ on t is sufficiently regular such that the integral is defined. In the present case strong continuity of $B(t)$ together with compactness of $K \; f(H)$ implies norm continuity of the integrand. Other weaker properties may suffice. We will mention this condition in the concrete applications below. Often $B(t)$ is norm continuous.

Proof. Observe that $\hat{P}(1)$ and $f(H)$ commute. Since $K \; f(H)$ is compact (uniformly for uniform families) it can be approximated in norm by a finite dimensional operator. Decay for each one dimensional operator then implies the decay (5.1). Thus it is sufficient to replace $K\ f(H)\ \hat{P}(1)$ by $(\Psi,\cdot)\Phi$ with $\Psi \in \mathrm{Ran}(\hat{P}(1)) = H^{\mathrm{cont}}(H)$, $\|\Psi\| = 1$, $\Phi \in H$. Then the square of the norm in (5.1) is

$$\left\| \frac{1}{T^2} \int_0^T dt \int_0^T ds\ B(t)\Phi\ (\Psi, e^{-iH(t-s)}\Psi)(\Phi, B^*(s)\cdot) \right\|$$

$$\leq M^2 \|\Phi\|^2\ \frac{1}{T^2} \int_0^T dt \int_0^T ds\ |(\Psi, e^{-iH(t-s)}\Psi)|$$

$$\leq \mathrm{const}\ \frac{1}{2T} \int_{-T}^{T} d\tau\ |(\Psi, e^{-iH\tau}\Psi)|$$

$$\leq \mathrm{const}\ \left\{ \frac{1}{2T} \int_{-T}^{T} d\tau\ |(\Psi, e^{-iH\tau}\Psi)|^2 \right\}^{1/2} .$$

We have used the Schwarz inequality in the last step. With the spectral theorem and $H = \int \lambda\ dE(\lambda)$ the scalar product reads $\int \exp(-i\lambda\tau)\ d\mu(\lambda)$ with the normalized continuous measure $\mu(\lambda) = (\Psi, E(\lambda)\Psi)$. Then

$$\{\ldots\} = \frac{1}{2T} \int_{-T}^{T} d\tau \iint e^{-i(\lambda-\lambda')\tau}\ d\mu(\lambda)\ d\mu(\lambda')$$

$$= \iint \frac{\sin(\lambda-\lambda')T}{(\lambda-\lambda')T}\ d\mu(\lambda)\ d\mu(\lambda')$$

$$\leq \frac{1}{|\delta T|} + \iint_{|\lambda-\lambda'|<\delta} d\mu(\lambda)\ d\mu(\lambda') .$$

By the continuity and boundedness of the measure μ the second summand is arbitrarily small for small δ. The first vanishes as $|T| \to \infty$ for any $\delta > 0$. □

We can apply this proposition to show local decay of scattering states uniformly in certain families of states. This has been called RAGE-theorem in [8]. In potential scattering let x denote the coordinate (or $x = X(d_1)$ in the present context), then $F(|x| < R) \in RC(1)$ for any $R < \infty$. A simple version is:

Corollary 5.3. Let $\Psi \in \mathcal{D}(H) \cap \mathcal{H}^{cont}(H)$, then

$$\frac{1}{T}\int_0^T dt\, \| F(|x|<R)\, e^{-iHt}\Psi \| \le f_R(T)\, \|(H-z)\Psi\| \tag{5.2}$$

where $f_R(T) \to 0$ as $|T| \to \infty$ for any $R < \infty$.

Proof. By the Schwarz inequality the square of the l.h.s. of (5.2) is bounded by

$$\frac{1}{T}\int_0^T dt\, \| F(|x|<R)\, e^{-iHt}\Psi \|^2$$

$$= \Big((H-z)\Psi, \frac{1}{T}\int_0^T dt\, e^{iHt}(H-\bar{z})^{-1}\, F(|x|<R)\,(H-z)^{-1}\, e^{-iHt}\,\hat{P}(1)\,(H-z)\Psi\Big)$$

$$\le \|(H-z)\Psi\|^2 \cdot \Big\| \frac{1}{T}\int_0^T dt\, e^{iHt}(H-\bar{z})^{-1}\, F(|x|<R)\,\hat{P}(1)\, e^{-iHt}\,(H-z)^{-1} \Big\|$$

$$=: \{f_R(T)\, \|(H-z)\Psi\|\}^2 .$$

By Prop. 5.1 $f_R(T)$ is independent of Ψ and it has the claimed decay. $B(t) = (H-\bar{z})^{-1}\exp(iHt)$ is norm-continuous. □

We will generalize these statements to multiparticle systems. Consider e.g. a three particle system where neither all three particles are bounded nor is a two body subsystem in a bound state. Then the separation between the particles in each pair increases in the long time average. We will show first a more general theorem and return to the physical discussion below. We begin with a generalization of Prop. 5.1. For d_1 (i.e. c_1 contains all particles, $H(c_1) = H$) they coincide. Here $B(t)$ need not respect the factorization of $\mathcal{H}$.

Lemma 5.4. For $\|B(t)\| \le M < \infty$, any $R < \infty$, any c_j

$$\lim_{|T|\to\infty} \Big\| \frac{1}{T}\int_0^T dt\, B(t)\, F(|X(c_j)|<R)\, e^{-iH(c_j)t} \times$$

$$\times\, [\mathbb{1} - P(c_j)]\, (H(c_j)-z)^{-1} \Big\| = 0 . \tag{5.3}$$

Proof. The square of the norm is bounded by a constant times

$$\frac{1}{T^2}\int_0^T dt \int_0^T ds \, \| K(c_j)\, e^{-iH(c_j)(t-s)}\, P^{cont}(H(c_j))\, K^*(c_j) \|$$

where

$$K(c_j) = F(|X(c_j)| < R)(H(c_j)-z)^{-1}$$

is compact as an operator in the internal subspace $\mathcal{H}(c_j)$ of the cluster c_j. There it can be replaced by a one dimensional operator. Since all operators are independent of the other variables the norm on $\mathcal{H}$ is the same as on the small space $\mathcal{H}(c_j)$. The rest of the proof is identical to that of Prop. 5.1. □

Lemma 5.5. Let $K \in RC(d_k)$, $\|B(t)\| \leq M < \infty \quad \forall t$.

$$\lim_{T\to\infty} \| \frac{1}{T}\int_0^T dt\, B(t)\, K[\mathbb{1} - P(d_k)] e^{-iH(d_k)t}(H-z)^{-1} \| = 0 . \tag{5.4}$$

The decay depends on K and M but not on $B(t)$.

Proof. For some denumeration of the clusters in d_k

$$\mathbb{1} - P(d_k) = \sum_{c_j \in d_k} [\mathbb{1} - P(c_j)] \prod_{i<j} P(c_i) .$$

We estimate each of the finitely many summands. For given c_j we can insert $F(|X(c_j)| < R)$ with small error when R is large. With the uniformly bounded

$$B'(t) = B(t)\, K \prod_{i<j} P(c_i)\; e^{-i(H(d_k)-H(c_j))t}$$

and $\|(H(c_j)-z)(H-z)^{-1}\| < \infty$ it remains to treat

$$\| \frac{1}{T}\int_0^T dt\, B'(t)\, F(|X(c_j)| < R)[\mathbb{1} - P(c_j)]\, e^{-iH(c_j)t}\, (H(c_j)-z)^{-1} \| .$$

This vanishes as $|T| \to \infty$ by Lemma 5.4. The uniformity in $B(t)$ is clear from the proof of Prop. 5.1 etc. □

With these preparations we are ready to show the main result of this section.

Theorem 5.6. Let $K \in RC(k)$, as defined at the beginning of Section IV. For any bounded function f which decays towards infinity and for any uniformly bounded family of operators $B(t)$ for which the integral is defined (e.g. norm continuous)

$$\lim_{|T|\to\infty} \| \frac{1}{T} \int_0^T dt\ B(t)\ K\ \hat{P}(k)\ e^{-iHt}\ f(H) \| = 0 . \qquad (5.5)$$

The same holds uniformly in u for uniform families $K_u \in RC(k)$.

The definition of $\hat{P}(k)$ is given in (2.17). For any decomposition with k or less clusters a state in the range of $\hat{P}(k)$ does not have all clusters bounded.

Proof. It is sufficient to show (5.5) with a rapid energy cutoff, e.g. $f(H) = (H-z)^{-1}$ or a power thereof: By the uniform boundedness of

$$C_T = \frac{1}{T} \int_0^T dt\ B(t)\ K\ \hat{P}(k)\ e^{-iHt} \qquad (5.6)$$

and

$$\|F(|H| > E)\ f(H)\| \to 0 \quad \text{as} \quad E \to \infty$$

it suffices to show e.g. for arbitrary fixed E

$$\|C_T\ (H-z)^{-3}\| \cdot \|(H-z)^3\ F(|H| \leq E)\| \to 0 \quad \text{as} \quad |T| \to \infty .$$

We proceed by induction over k . The start $k = 1$ is covered by Prop. 5.1. Due to (4.2) we have to treat only $K \in RC(d_k)$ for fixed d_k . We decompose $\hat{P}(k)$ accordingly.

$$\hat{P}(k) = [\mathbb{1} - P(d_k)]\ \hat{P}(k-1) - \sum_{d_k' \neq d_k} P(d_k')\ \hat{P}(k-1) . \qquad (5.7)$$

By Lemma 4.7 $K\ P(d_k') \in RC(k-1)$ and the corresponding terms are covered by the induction hypothesis. With Corollary 4.4 for $s = 0$ we obtain

$$\| \frac{1}{T} \int_0^T dt\ B(t)\ K\ [\mathbb{1} - P(d_k)]\ \hat{P}(k-1)\ e^{-iHt}\ (H-z)^{-3}$$

$$- \frac{1}{T} \int_0^T dt\ B(t)\ K\ [\mathbb{1} - P(d_k)]\ (H-z)^{-2}\ \hat{P}(k-1)\ e^{-iHt}\ (H-z)^{-1} \|$$

$$\leq \sum_{l=1}^{k-2} \left\| \frac{1}{T} \int_0^T dt\ B_l(t)\ K_l\ \hat{P}(l)\ e^{-iHt}\ (H-z)^{-1} \right\|$$

$$\to 0 \quad \text{as} \quad |T| \to \infty \tag{5.8}$$

by induction hypothesis and $K_l \in RC(l)$.

For given time length $0 < S < \infty$ we introduce the sawtooth function $s(t)$

$$s(t) := t - mS \quad \text{if} \quad mS \leq t < (m+1)S \ . \tag{5.9}$$

Then $0 \leq s(t) < S \quad \forall\ t \in \mathbb{R}$ and $t - s(t)$ is a step function which is constant on intervals of length S . The value of S will be chosen later.

The negative term in (5.8) is bounded by

$$\left\| \frac{1}{T} \int_0^T dt\ B(t)\ K\ [\mathbb{1} - P(d_k)]\ e^{-iH(d_k)s(t)}\ e^{iHs(t)}\ (H-z)^{-2}\ \hat{P}(k-1)\ \times \right.$$

$$\left. \times\ e^{-iHt}\ (H-z)^{-1} \right\|$$

$$+ \left\| \frac{1}{T} \int_0^T dt\ B(t)\ K'\ \{\mathbb{1} - e^{-iH(d_k)s(t)}\ e^{iHs(t)}\}\ (H-z)^{-2}\ \hat{P}(k-1)\ \times \right.$$

$$\left. \times\ e^{-iHt}\ (H-z)^{-1} \right\| \ . \tag{5.10}$$

Here $K' = K\ [\mathbb{1} - P(d_k)] \in RC(d_k)$ and by Cor. 4.5

$$K'\ \{\mathbb{1} - e^{-iH(d_k)s(t)}\ e^{iHs(t)}\}\ (H-z)^{-2} \in RC(k-1)$$

uniformly in $t \in \mathbb{R}$ for any finite S . Consequently the second term in (5.10) vanishes as $|T| \to \infty$ by induction hypothesis. We bound the first summand by

$$\left\| \frac{1}{T} \int_0^T dt\ B(t)\ K\ [\mathbb{1} - P(d_k)]\ e^{-iH(d_k)s(t)}\ (H-z)^{-1}\ \times \right.$$

$$\left. \times\ \hat{P}(k-1)\ e^{-iH(t-s(t))}\ (H-z)^{-2} \right\|$$

$$+ \sum_{l=1}^{k-2} \left\| \frac{1}{T} \int_0^T dt\ B_l(t)\ K_l(s(t))\ \hat{P}(l)\ e^{-iHt}\ (H-z)^{-1} \right\| \ . \tag{5.11}$$

By Corollary 4.4 the $K_l(s(t))$ are uniformly clustered for $t \in \mathbb{R}$ and the sum over l vanishes as $|T| \to \infty$ by induction for any finite S. In the first term we can carry out the integral over intervals of length S where $t - s(t)$ is constant. With $|T| = nS$

$$\frac{1}{T}\int_0^T = \frac{1}{n}\sum_{m=0}^{n-1} \frac{1}{S}\int_{mS}^{(m+1)S}$$

and we obtain the bound

$$\max_m \left\| \frac{1}{S}\int_0^S ds\; B(mS+s)\; K\, [\mathbb{1} - P(d_k)]\, e^{-iH(d_k)s}\, (H-z)^{-1} \right\| \times$$
$$\times \; \|\hat{P}(k-1)\,(H-z)^{-2}\| \;. \tag{5.12}$$

For given $\varepsilon > 0$ we choose by Lemma 5.5 the time S large enough such that (5.12) is smaller than $\varepsilon/4$. For times which are not integer multiples of S we get an additional term const S/T. This implies the bound $\varepsilon/2$ for that term with arbitrary $|T| \geq T_1 \geq S$. Then for large enough $|T|$ all the finitely many other terms are also smaller than $\varepsilon/2$ as mentioned above. □

As a heuristic illustration why the above proof works we will give the argument once more for a three body system, $k = 2$, and $K = F(|x_1 - x_2| < r)$.

$$\hat{P}(2) = [\mathbb{1} - P(1,2)]\,\hat{P}(1) - \{P(1,3) + P(2,3)\}\,\hat{P}(1) \;.$$

Clearly $K\,\{P(1,3) + P(2,3)\}$ is relatively compact on $\mathcal{H}$ and the time average vanishes on the continuous spectral subspace Ran $\hat{P}(1)$ by the well known Prop. 5.1. If y_3 denotes the position of the third particle relative to the center of mass of the pair (1,2) we decompose

$$K\,[\mathbb{1} - P(1,2)] = K\,[\mathbb{1} - P(1,2)]\,F(|y_3| < R)$$
$$+ K\,[\mathbb{1} - P(1,2)]\,F(|y_3| > R) \;.$$

The first summand is again relatively compact for any $R < \infty$ and its time average vanishes as above. For the time evolution where those potentials are omitted which couple the third particle to the pair we apply Prop. 5.1 for the subsystem of the pair and obtain

$$\| \frac{1}{S} \int_0^S ds\ e^{iHs}\ F(|x_1-x_2| < r)[\mathbb{1} - P(1,2)]\ e^{-iH(1,2-3)s}(H(1,2)-z)^{-1} \| < \varepsilon \ .$$

S depends on r only. For this S we can find an $R < \infty$ such that

$$\sup_{0\le s\le S} \| F(|x_1-x_2|<r)[\mathbb{1}-P(1,2)]\ F(|y_3|>R)\ \{e^{-iH(1,2-3)s} - e^{-iHs}\}(H-z)^{-1} \| < \varepsilon \ .$$

Thus for these parameters

$$\| \frac{1}{S} \int_0^S ds\ e^{iHs}\ F(|x_1-x_2|<r)[\mathbb{1}-P(1,2)]\ F(|y_3|>R)\ e^{-iHs}(H-z)^{-1} \| < \text{const}\ \varepsilon \ .$$

Longer time averages are bounded by this.

For higher numbers of clusters one uses this approach to reduce it by one. The decay estimates are expressed as cluster relative compactness properties in our proof.

Instead of the usual time average one can use weighted averages as well.

Corollary 5.7. Under the assumptions of Theorem 5.6

$$\lim_{|T|\to\infty} \| \frac{1}{T^2} \int_0^T 2t\,dt\ B(t)\ K\ \hat{P}(k)\ e^{-iHt}\ f(H) \| = 0 \ .$$

Proof. By partial integration

$$\frac{1}{T^2} \int_0^T 2t\,dt\ C(t) = \frac{1}{T^2}\ 2T \int_0^T ds\ C(s) - \frac{1}{T} \int_0^T \frac{2t}{T} dt\ \frac{1}{t} \int_0^t ds\ C(s) \ . \qquad (5.13)$$

Both terms on the right hand side decay if the ordinary time average decays. □

We will use our theorem in the sections below to estimate the effects of intercluster potentials on the motion. Here we state local decay properties which follow directly. We have not used so far the condition on the derivative of the potentials. All statements follow from the asymptotic decay (1.2) of the pair potentials alone! One can insert into (5.5) the prototype $K = \prod F(|X(c_j)| < R)$ for some d_k

and R. More interesting is the following statement about separation of clusters.

Corollary 5.8. Let $|y^\alpha|$ denote the separation of the centers of mass of any two clusters in d_k and choose $R < \infty$. Then for $|T| \to \infty$

(a)
$$\sup_{s\in\mathbb{R}} \frac{1}{T} \int_s^{s+T} dt \; \| F(|y^\alpha| < R) \; \tilde{P}(d_k) \; e^{-iHt} \Psi \| \to 0 \quad . \tag{5.14}$$

(b) There is a sequence $\tau_n \to \infty$ (or $\to -\infty$) such that

$$\lim_{n\to\infty} \| F(|y^\alpha| < R) \; \tilde{P}(d_k) \; e^{-iH\tau_n} \Psi \| = 0 \; . \tag{5.15}$$

Proof. (b) is an obvious consequence of (a). This follows from $F(|y^\alpha| < R)\, P(d_k) = K \in RC(k-1)$ for any R and $F(|y^\alpha| < R)\, \tilde{P}(d_k) = K\, \hat{P}(k-1)$. We get the integral inside the norm like in the proof of Corollary 5.3. □

VI. The First Reduction, Summation over Decompositions.

The operators in (3.1) - (3.4) are all bounded, therefore it is sufficient to treat an arbitrary vector Ψ from the dense set

$$\mathcal{D} = \mathcal{D}(H) \cap \mathcal{D}(|Y|^2) \cap \mathcal{D}(|Y|^2 H) \quad . \tag{6.1}$$

It is known to be time-translation invariant

$$e^{-iHt} \mathcal{D} = \mathcal{D} \qquad \forall \; t \in \mathbb{R} \;, \tag{6.2}$$

and by the decay assumptions on eigenfunctions (2.16)

$$P(d_k)\; \mathcal{D} \subset \mathcal{D} \;, \quad \hat{P}(k)\; \mathcal{D} \subset \mathcal{D} \qquad \forall \; k \quad . \tag{6.3}$$

We omit the proof of these statements, for $N = 3$ see [3] and see [4] for general N.

We will show first that for a fixed $\Psi \in \mathcal{D}$, any $\varepsilon > 0$

$$\sup_{s\in\mathbb{R}} \frac{1}{T} \int_s^{s+T} dt \; \| [g(H) - g(H(d_k))] \; \tilde{P}(d_k) \; e^{-iHt} \Psi \| < \varepsilon \tag{6.4}$$

for all $|T| \geq T(\varepsilon)$. As in the proof of Corollary 5.3 by the Schwarz inequality it is sufficient to show decay of

$$\frac{1}{T}\int_s^{s+T} dt \ \| [g(H) - g(H(d_k))] \ \tilde{P}(d_k) \ e^{-iHt} \Psi \|^2 =$$

$$= \left(e^{-iHs}(H-z)\Psi, \frac{1}{T}\int_0^T dt \ (H-z)^{-1} e^{iHt} \tilde{P}(d_k)^* [\ldots]^2 \ \tilde{P}(d_k) e^{-iHt} (H-z)^{-1} \times \right.$$

$$\left. \times \ e^{-iHs} (H-z)\Psi\right)$$

$$\leq \ \|(H-z)\Psi\|^2 \ \| \frac{1}{T}\int_0^T dt \ B(t)[g(H)-g(H(d_k))] \ P(d_k) \ \hat{P}(k-1) \ e^{-iHt} (H-z)^{-1} \| \quad (6.5)$$

where $B(t) = (H-\bar{z})^{-1} \exp(iHt) \tilde{P}(d_k)^* [g(H)-g(H(d_k))]^*$. The integrand is norm continuous in t and therefore the integral exists. By Lemma 4.3 (a) $[g(H)-g(H(d_k))] \ P(d_k) \in RC(k-1)$. Thus Theorem 5.6 implies decay in T .

The equivalence of (3.2) and (3.3) follows easily from the fact that

$$[Y(d_k)/\tau, \ Q(d_k)]$$

is a multiple of the identity which decays as $\tau \to \infty$. With the Baker-Campbell-Hausdorff formula the asymptotic equality follows for exponential functions and by Fourier representation for arbitrary f_k with integrable Fourier transform like e.g. $f_k \in C_0^\infty$ (which is sufficient for applications). Moreover by standard arguments it is sufficient to show for $\Psi \in \mathcal{D}$ decay of

$$\frac{1}{2} \left\| \left\{ \frac{Y(d_k)}{t} - Q(d_k) \right\} \tilde{P}(d_k) \ e^{-iHt} \Psi \right\|^2 =$$

$$= \left(\Psi, e^{iHt} \tilde{P}(d_k)^* \left\{ \frac{|Y(d_k)|^2}{2t^2} - \frac{D(d_k)}{t} + T(d_k) \right\} \tilde{P}(d_k) \quad e^{-iHt} \Psi\right) \quad (6.6)$$

for $t = \tau_n$ suitably chosen. We have absorbed the mass factors (cf.(2.3)) in our definition of $\|\cdot\|^2$ for vectors. See e.g. Lemma 2.6 in [3] for the explicit calculations in a similar context. The operators are defined in Section II and $\mathcal{D}$ is contained in their domains. $P(d_k)$ commutes with $Y(d_k)$ and $D(d_k)$ and thus can be removed from the left. Using the simplifying assumption (2.16) it is very easy to estimate

$$\left\| \left\{ \frac{|Y(d_k)|^2}{t^2} - \frac{|Y|^2}{t^2} \right\} \widetilde{P}(d_k) \right\| \leq \sum_{c_j \in d_k} \| |X(c_j)|^2 P(c_j) \| / t^2 \to 0 \qquad (6.7)$$

as $|t| \to \infty$. Similarly

$$\left\| \left\{ \frac{D(d_k)}{t} - \frac{D}{t} \right\} \widetilde{P}(d_k) \right\| \leq \text{const} / |t| \to 0 \quad . \qquad (6.8)$$

Thus in the Y- and D- terms we can move iteratively all bound state projections from left to right and get for $|t| \to \infty$

$$\left\| \widetilde{P}(d_k)^* \left\{ \frac{|Y(d_k)|^2}{2\,t^2} - \frac{D(d_k)}{t} \right\} \widetilde{P}(d_k) - \left\{ \frac{|Y|^2}{2\,t^2} - \frac{D}{t} \right\} \hat{P}(k-1)^* \widetilde{P}(d_k) \right\| \to 0 \ . \quad (6.9)$$

Since by Corollary 4.8 (a) $[\mathbb{1} - \hat{P}(k-1)^*]\, P(d_k) = K \in RC(k-1)$ we take time averages to eliminate $\hat{P}(k-1)^*$.

$$\left| \frac{1}{T} \int_s^{s+T} dt \left(\Psi, e^{iHt} \left\{ \frac{|Y|^2}{2\,t^2} - \frac{D}{t} \right\} [\mathbb{1} - \hat{P}(k-1)^*]\, \widetilde{P}(d_k)\, e^{-iHt}\, \Psi \right) \right|$$

$$\leq \| (|Y|^2 + \mathbb{1})(H-z)\Psi \| \; \| (H-z)\Psi \| \cdot \left\| \frac{1}{T} \int_s^{s+T} dt\; B(t)\, K\, \hat{P}(k-1)\, e^{-iHt} (H-z)^{-1} \right\| , \quad (6.10)$$

where

$$B(t) := (|Y|^2 + \mathbb{1})^{-1} (H-z)^{-1} e^{iHt} \left\{ \frac{Y^2}{2\,t^2} - \frac{D}{t} \right\}$$

is uniformly bounded and norm-continuous. The proof of this is usually an intermediate step for (6.2) and we omit it as well, see [4]. By Theorem 5.6 the expression (6.10) decays as $T \to \infty$ uniformly in s and the previous error (6.9) is small for all large s. For negative times we would take the integral from $-T+s$ to s with T and $-s$ large. From now on we will treat only positive times.

The decay of (6.6) for all d_k and k follows from the decay of the sum. With our reductions we have eliminated the d_k-dependence of the Y- and D- terms and we can exploit that $\sum \sum \widetilde{P}(d_k) = \mathbb{1}$. This innocent looking trick is crucial to avoid in the estimates below like (7.2) commutators of H with $P(d_k)$. It seems that their control would require much more subtle estimates than our elementary compactness arguments.

With the above reductions our next goal is to estimate for large s and T:

$$\frac{1}{T}\int_{s}^{s+T} dt \left(\Psi, e^{iHt}\left\{\frac{|Y|^2}{2t^2} - \frac{D}{t} + \sum_{k}\sum_{d_k} \tilde{P}(d_k)^* T(d_k)\, \tilde{P}(d_k)\right\} e^{-iHt}\,\Psi\right) \quad . \tag{6.11}$$

This will be done in the next section.

VII. The Use of Kinematics, Completion of the Proof.

For $|Y|^2(t) := \exp(iHt)\,|Y|^2 \exp(-iHt)$ the derivative is defined as a quadratic form between vectors in $\mathcal{D}$ by the usual commutator formula. Since $|Y|^2$ commutes with the potentials

$$\frac{1}{2}\, i[H, |Y|^2] = \frac{1}{2}\, i[H_0, |Y|^2] = D \quad . \tag{7.1}$$

Thus

$$\frac{1}{2t^2}\left(|Y|^2(t) - |Y|^2\right) = \frac{1}{t^2}\int_0^t d\tau\, \frac{d}{d\tau}\, \frac{1}{2}\,|Y|^2(\tau)$$

$$= \frac{1}{t^2}\int_0^t d\tau\, D(\tau) = \frac{1}{t^2}\int_0^t d\tau \int_0^\tau \dot{D}(s)\, ds + \frac{D(0)}{t}$$

where $\dot{D}(s) = d\,D(s)/ds = i[H,D](s)$ is also defined as a quadratic form. By partial integration we get

$$\frac{|Y|(t) - |Y|^2}{2t^2} - \frac{D(t)}{t}$$

$$= \frac{1}{t^2}\int_0^t d\tau \int_0^\tau ds\, \dot{D}(s) - \frac{1}{t}\int_0^t ds\, \dot{D}(s)$$

$$= -\frac{1}{t^2}\int_0^t \tau\, d\tau\, \dot{D}(\tau) \quad . \tag{7.2}$$

Clearly $(\Psi, |Y|^2\Psi)/2t^2 \to 0$ and we can insert (7.2) in the integrand of (6.11). Reintroducing the decomposition of the identity we obtain terms of the form

$$\frac{1}{t^2}\int_0^t \tau\, d\tau\, (\Psi, e^{iH\tau}\, i[H,D]\, \tilde{P}(d_k)\, e^{-iH\tau}\,\Psi)$$

$$- (\Psi, e^{iHt}\, \tilde{P}(d_k)^*\, T(d_k)\, \tilde{P}(d_k)\, e^{-iHt}\,\Psi) \quad . \tag{7.3}$$

The commutator equals

$$
\begin{aligned}
& i[H,D]\ P(d_k) \\
& = i[T(d_k),D]\ P(d_k) + \sum_{c_j \in d_k} i[H(c_j),D]\ P(d_k) \\
& \quad + \sum_{\alpha \notin d_k} i[V^\alpha,D]\ P(d_k) \\
& = 2\ T(d_k)\ P(d_k) + \sum_{c_j \in d_k} i[H(c_j),D(c_j)]\ P(d_k) \\
& \quad - \sum_{\alpha \notin d_k} x^\alpha \cdot (\nabla V^\alpha)(x^\alpha)\ P(d_k)\ .
\end{aligned} \tag{7.4}
$$

By Corollary 4.8 (b) the last terms are in $RC(k-1)$ and the weighted time average of these terms decays as $|t| \to \infty$ by Corollary 5.7. We use the calculation (5.13) also for the contribution from the second term on the r.h.s. of (7.4), i.e. we will show asymptotic decay of the ordinary time average and conclude decay of the weighted average. By Lemma 4.9 (a) $i[H(c_j),D(c_j)]\ P(d_k) \in RC(d_k)$. Exactly as in the proof of Theorem 5.6 we introduce the sawtooth approximation $\exp(-iH(d_k)s)$ of $\exp(-iHs)$ on time intervals of arbitrary long but fixed length S . We do it on the left as well which is possible by Lemma 4.9 (b). Thus we obtain as an approximation for arbitrary fixed S and long $t \geq T_2 \geq S$

$$
\begin{aligned}
& \Big| \frac{1}{t} \int_0^t d\tau\ (\Psi, e^{iH(\tau - s(\tau))}\ e^{iH(d_k)s(\tau)}\ i[H(c_j),D(c_j)]\ P(d_k)\ e^{-iH(d_k)s(\tau)}\ \times \\
& \qquad\qquad \times\ \hat{P}(k-1) e^{-iH(\tau - s(\tau))}\ \Psi) \Big| \\
& \leq 2\ \Big\| \frac{1}{S} \int_0^S ds\ e^{iH(c_j)s}\ i[H(c_j),D(c_j)]\ P(c_j)\ e^{-iH(c_j)s} \Big\|\ \|\hat{P}(k-1)\|\ \|\Psi\|^2\ .
\end{aligned} \tag{7.5}
$$

In the last inequality we have used that $\{H(d_k) - H(c_j)\}$ commutes with $i[H(c_j),D(c_j)]\ P(d_k)$ and the argument following (5.12). The integrand is norm continuous in s . The operators depend only on the internal coordinates of the cluster c_j and the norm can be evaluated on the subspace $\mathcal{H}(c_j)$. There $P(c_j) = \sum P_l$ with $H(c_j)\ P_l = E_l\ P_l$ is a finite sum of one-dimensional projections. Then

$$\| \frac{1}{S} \int_0^S ds\, e^{iH(c_j)s}\, i[H(c_j), D(c_j)]\, P_1\, e^{-iH(c_j)s} \|$$

$$= \| \frac{1}{S} \int_0^S ds\, i(H(c_j)-E_1)\, e^{i(H(c_j)-E_1)s}\, D(c_j)\, P_1 \|$$

$$= \| \frac{1}{S} \{ e^{i(H(c_j)-E_1)S} - \mathbb{1} \}\, D(c_j)\, P_1 \| \le \frac{1}{S}\, 2\, \| D(c_j)\, P_1 \| \ . \tag{7.6}$$

For a given $\varepsilon > 0$ we can determine a suitable S and deduce with the above estimates

$$\limsup_{t\to\infty} \left| \frac{1}{t^2} \int_0^t \tau\, d\tau\, (\Psi, e^{iH\tau} \sum_{c_j \in d_k} i[H(c_j), D]\, \widetilde{P}(d_k)\, e^{-iH\tau}\Psi) \right| < \varepsilon \ . \tag{7.7}$$

The only remaining term from (7.4) is $2\,T(d_k)\,P(d_k)$. By Lemma 4.10 $(H-z)^{-1}\, [\mathbb{1} - \hat{P}(k-1)^*] T(d_k)\, P(d_k) \in RC(k-1)$ and we can insert with Corollary 5.7 a factor $\widetilde{P}(d_k)^*$ left of $T(d_k)$ in the long time weighted average. (c.f. the inverse procedure in Section VI.) Thus for the estimate of (7.3) it remains to study

$$\frac{1}{t^2} \int_0^t 2\,\tau\, d\tau\, G(\tau) - G(t)$$

where

$$G(t) := \sum_k \sum_{d_k} (\Psi, e^{iHt}\, \widetilde{P}(d_k)^*\, T(d_k)\, \widetilde{P}(d_k)\, e^{-iHt}\Psi) \ge 0 \tag{7.8}$$

is a continuous uniformly bounded function. Returning to (6.11) we have to control

$$\frac{1}{T} \int_s^{s+T} dt\, \{ \frac{1}{t^2} \int_0^t 2\,\tau\, d\tau\, G(\tau) - G(t) \} \ . \tag{7.9}$$

A bounded continuous real valued function cannot stay away from its average forever. Therefore one can show that for any T there is a sequence s'_n such that (7.9) vanishes as $s'_n \to \infty$. For a proof see Lemma 8.15 in [3]. All our approximations above become better for T and s larger. Thus we have shown: For any increasing sequence $T_n \to \infty$ there is a sequence $s_n \to \infty$ such that

$$\frac{1}{T_n} \int_{s_n}^{s_n+T_n} dt \sum_k \sum_{d_k} \{ \| [g(H)-g(H(d_k))] \tilde{P}(d_k)\, e^{-iHt} \Psi \|$$

$$+ \frac{1}{2} \| \{ \frac{Y(d_k)}{t} - Q(d_k) \} \tilde{P}(d_k)\, e^{-iHt} \Psi \| \}$$

$$\to 0 \quad \text{as} \quad n \to \infty \;.$$

In particular there must be $\tau_n \in [s_n, s_n+T_n]$ where the integrand vanishes as $n \to \infty$. This completes the proof of Theorem 3.1.

We have seen in the proof that the crucial calculations are pure kinematics, namely $i[H,|Y|^2] \equiv i[H_0,|Y|^2] = 2D$ and $i[T(d_k),D] = 2T(d_k)$. By looking at $Y/t - Q$ rather than $Y - Qt$ we had to estimate only time averages. All perturbations caused by interaction terms from intercluster potentials were eliminated using Theorem 5.6. The mathematical machinery agrees perfectly with the physical intuition. On the other hand the forces within the clusters are crucial to keep them bounded. This allowed to treat a k-cluster system essentially as a k-body system. The subsystem Hamiltonians have not been decomposed and the estimates (7.5) - (7.7) are nothing but a variant of the virial theorem. Summing up we can say that our proof used only kinematical relations and iterations of the "soft" Proposition 5.1. One had to look at suitably chosen combinations, e.g. replacing a $\nu(k-1)$-dimensional vector equation by one involving three scalars (6.6) or the summing over decompositions. That the proof is nevertheless so long follows from the complexity of N-body systems and the numerous approximations. It is elementary.

If singular short-range potentials are added then one has to use some regularizations by resolvents and treat the commutator with D termwise. The assumption about decay of eigenfunctions can easily be avoided by introducing an additional spatial cutoff.

When the subsystems have infinitely many bound states one has to fix first an error bound ε and choose for each k a number $N(\varepsilon,k)$ of bound states. The remaining weakly bounded ones are treated together with the scattering states. Then all compactness arguments remain valid. One has to estimate some additional terms and care is needed to avoid a circularity in the choice of the $N(\varepsilon,k)$'s. Since bound state projections for subsystems are not time invariant we believe that the above splitting is more convenient in scattering theory than the usual into continuous and point-spectral subspaces. The details of the above generalizations will be given in [4].

References.

1. V. Enss: Geometric methods in spectral and scattering theory of Schrödinger operators, in: Rigorous Atomic and Molecular Physics; G. Velo and A.S. Wightman eds., Plenum, New York 1981, pp. 1 - 69.

2. V. Enss: Asymptotic observables on scattering states, Commun. Math. Phys. 89, 245 - 268 (1983).

3. V. Enss: Quantum scattering theory for two- and three-body systems with potentials of short and long range, in: Schrödinger Operators, S. Graffi ed., Springer L N Math. 1159, Berlin 1985, p. 39 - 176.

4. V. Enss: (a) Separation of subsystems and clustered operators for multiparticle quantum systems, preprint Nr.213, Mathematics, Freie Universität Berlin;
(b) Observables and asymptotic phase space localization of N-body quantum scattering states, in preparation.

5. R. Froese and I. Herbst: Exponential bounds and absence of positive eigenvalues for N-body Schrödinger operators, Commun. Math. Phys. 87, 429 - 447 (1982).

6. PL. Muthuramalingam and K. B. Sinha: Asymptotic completeness in long range scattering II, Ann. scient. Ec. Norm. Sup. 18, 57 - 87 (1985).

7. P. A. Perry, I. M. Sigal, and B. Simon: Spectral analysis of N-body Schrödinger operators, Ann. Math. 114, 519 - 567 (1981).

8. M. Reed, B. Simon: Methods of Modern Mathematical Physics, III. Scattering Theory, Academic Press, Now York 1979.

9. K. B. Sinha and PL. Muthuramalingam: Asymptotic evolution of certain observables and completeness in Coulomb scattering I, J. Func. Anal. 55, 323 - 343 (1984).

10. B. Thaller and V. Enss: Asymptotic observables and Coulomb scattering for the Dirac equation, preprint Nr.198, Mathematik, Freie Universität Berlin; to appear in Ann. Inst. H. Poincaré Sec. A.

SCATTERING THEORY FOR ONE-DIMENSIONAL SYSTEMS WITH NONTRIVIAL SPATIAL ASYMPTOTICS

F. Gesztesy
Institute for Theoretical Physics
University of Graz
A-8010 Graz, Austria

1. Introduction

In the first part of this paper we consider Schrödinger operators $H=-\frac{d^2}{dx^2}+V$ in $L^2(\mathbb{R})$ with potentials V having nontrivial (i.e. in general nonzero) spatial asymptotics. In particular we are interested in the cases where

a) $V(x) \xrightarrow[x\to\pm\infty]{} V_\pm \in \mathbb{R}$, $V_- \leq V_+$.

b) $V(x) \xrightarrow[x\to-\infty]{} V_- \in \mathbb{R}$, $V(x) \xrightarrow[x\to+\infty]{} +\infty$.

c) $V(x)$ is periodic on $(0,\pm\infty)$ with in general different periods $a_\pm > 0$.

Our main concern is to study the possibility of total reflection of a particle on the potential V in certain energy regimes. It turns out that these energy regimes are given precisely by those intervals in the absolutely continuous spectrum $\sigma_{ac}(H)$ of H where H has simple spectrum. In the remaining part of $\sigma_{ac}(H)$ where the spectral multiplicity of H equals two, transmission of the particle from the left and right occurs with a nonzero probability. These results have been derived by Davies and Simon [16] using abstract methods together with a natural decomposition of the projection onto the absolutely continuous subspace associated with H into states running towards $-\infty$ (resp. $+\infty$) as $t\to\pm\infty$. Our approach will be somewhat complementary and relies entirely on stationary scattering methods. In particular we shall use elementary concepts such as Jost solutions and Wronski determinants to derive the connection between total reflection and spectral multiplicity one mentioned above. In fact it will turn out that under rather general assumptions on V the number of linearly independent bounded Jost solutions on an open energy interval $I \subset \sigma_{ac}(H)$ coincides with the spectral multiplicity of H on I.

In Sect. 2 we treat case a) mentioned above and show that H has simple spectrum in (V_-,V_+) (in case $V_-<V_+$) and spectral multiplicity two in (V_+,∞). Following in part Cohen and Kappeler [15] we consider basic properties of reflection and transmission coefficients by studying the underlying Wronski determinants in some details. At the end of this section we consider relative scattering between asymptotically similar potentials V_j (i.e. $V_j(x) \xrightarrow[x\to\pm\infty]{} V_\pm$), j=1,2. In particular we derive the connection between the relative transmission coefficient and the Fredholm determinant between $H_1=-\frac{d^2}{dx^2}+V_1$ and $H_2=-\frac{d^2}{dx^2}+V_2$ generalizing a proof of Newton [36].

Case b) is treated in Sect. 3. It turns out that H has simple absolutely continuous spectrum in (V_-,∞) and hence total reflection occurs for all energies $\lambda > V_-$.

Sect. 4 corresponds to case c) above and treats systems composed of two half crystals. There the intersection of the spectra of the pure crystals yields the energies where H has spectral multiplicity two whereas the spectrum of H is simple in the remaining part of the union of the pure crystal spectra.

The second part of this paper consists of Sect. 5 and treats applications in supersymmetric quantum mechanics. We first give a short introduction to the connection of Krein's spectral shift function between the pair of "bosonic" and "fermionic" Hamiltonians and Witten's (regularized) index resp. the anomaly based on [5] and [22]. Exploiting the methods for relative scattering described at the end of Sect. 2 together with results of [17] we explicitly calculate Krein's spectral shift function, Witten's (regularized) index and the corresponding anomaly for supersymmetric systems subordinated to Sects. 2 or 3. We also consider similar problems on the half line $(0,\infty)$.

2. The case $V(x) \to V_\pm \in \mathbb{R}$ as $x\to\pm\infty$

In this section we use assumptions of the type

H(α). Let $V \in L^1_{loc}(\mathbb{R})$ be real-valued, $V_\pm \in \mathbb{R}$, $V_- \leq V_+$ and

$$\int_{-\infty}^{0} dx(1+|x|^\alpha)|V(x)-V_-| + \int_0^\infty dx(1+|x|^\alpha)|V(x)-V_+| < \infty$$

for some $\alpha \geq 0$.

Unless otherwise stated we always assume H(0) in this section. Given H(0) we introduce in $L^2(\mathbb{R})$ the form sum

$$H=-\frac{d^2}{dx^2}\mp V, \quad \mathcal{D}(-\frac{d^2}{dx^2})=H^{2,2}(\mathbb{R}). \tag{2.1}$$

Our main tools to study H are based on corresponding Jost solutions $f_\pm(z,x)$ defined as unique solutions of

$$f_\pm(z,x)=e^{\pm ik_\pm x}-\int_x^{\pm\infty} dx' k_\pm^{-1}\sin[k_\pm(x-x')][V(x')-V_\pm]f_\pm(z,x'),$$
$$z\in\mathbb{C},\ x\in\mathbb{R} \tag{2.2}$$

where

$$k_\pm(z)=(z-V_\pm)^{1/2},\ \mathrm{Im}k_\pm(z)\geq 0,\ z\in\mathbb{C}. \tag{2.3}$$

By standard considerations (cf. e.g. [14], [37]) H(1) implies that $f_\pm(z,x)$ are analytic w.r. to $z\in\mathbb{C}\setminus[V_\pm,\infty)$. In particular one obtains the estimates

$$|f_\pm(z,x)|\leq c_{1\pm}(z)e^{\mp(\mathrm{Im}k_\pm)x},\ |f'_\pm(z,x)|\leq\tilde{c}_{1\pm}(z)e^{\mp(\mathrm{Im}k_\pm)x},$$
$$x \gtrless 0,\ z\in\mathbb{C}\setminus\{V_\pm\} \tag{2.4}$$

assuming H(0) ($c_{1\pm}(z)$ stays bounded as $|z|\to\infty$) and

$$|\dot{f}_\pm(z,x)|\leq c_{2\pm}(z)|x|e^{\mp(\mathrm{Im}k_\pm)x},\ |\dot{f}'_\pm(z,x)|\leq\tilde{c}_{2\pm}(z)|x|e^{\mp(\mathrm{Im}k_\pm)x},$$
$$x \gtrless 0,\ z\in\mathbb{C}\setminus\{V_\pm\} \tag{2.5}$$

assuming H(1). (Here "prime" denotes $\partial/\partial x$, "dot" abbreviates $\partial/\partial z$ and each equation holds separately for "+" ($x\geq 0$) and "-" ($x\leq 0$)). Assuming H(1) we get

$$|f_-(z,x)|\leq c_3 e^{(\mathrm{Im}k_-)x},\quad x\leq 0,\ z\in\mathbb{C}, \tag{2.6}$$

and assuming H(2) we obtain

$$|f_-(V_-,x)-1|\leq c_4(1+|x|)^{-1},\ x\leq 0, \tag{2.7}$$

$$|(\partial/\partial k_-)f_-(z,x)|\leq c_5|x|e^{(\mathrm{Im}k_-)x},\quad x\leq 0,\ z\in\mathbb{C}. \tag{2.8}$$

We have

$$f_\pm(z,x)=c_\mp(z)f_\mp(z,x)+d_\mp(z)\overline{f_\mp(z,x)},\ z\in\mathbb{C}\setminus\{V_\pm\} \tag{2.9}$$

where

$$2ik_+(z)d_+(z)=2ik_-(z)d_-(z)=W(f_-(z),f_+(z)),$$
$$-2ik_+(z)c_+(z)=W(f_-(z),\overline{f_+(z)}),$$
$$-2ik_-(z)c_-(z)=W(\overline{f_-(z)},f_+(z));\ z\in\mathbb{C}\setminus\{V_\pm\} \tag{2.10}$$

and

$$W(F,G)_x = F(x)G'(x) - F'(x)G(x) \tag{2.11}$$

denotes the Wronskian of F and G. We start with

Lemma 2.1. Assume H(0) and abbreviate $W(z)=W(f_-(z),f_+(z))$, $z\in\mathbb{C}$. Then $W(z)\neq 0$ for $z\in\mathbb{C}\backslash(-\infty,V_-]$. If in addition H(1) holds then $W\in C^o(\mathbb{R})$. Moreover W is analytic w.r. to $z\in\mathbb{C}\backslash[V_-,\infty)$.

Proof. It suffices to sketch a proof of the first assertion. The identity

$$W(F,G)_x W(\phi,\psi)_x = W(F,\psi)_x W(\phi,G)_x - W(F,\phi)_x W(\psi,G)_x \tag{2.12}$$

implies for $\lambda>V_+$

$$W(f_+(\lambda),f_-(\lambda))W(\overline{f_+(\lambda)},\overline{f_-(\lambda)}) = |W(f_+(\lambda),\overline{f_-(\lambda)})|^2 + 4k_+k_- \geq 4k_+k_- > 0 \tag{2.13}$$

since

$$W(\overline{f_\mp(z)},f_\mp(z)) = \begin{cases} \mp 2ik_\mp, & z\in\mathbb{C}\backslash(-\infty,V_\mp] \\ 0, & z\leq V_\mp . \end{cases} \tag{2.14}$$

Next assume that there is a $\lambda_o\in(V_-,V_+]$ such that $W(\lambda_o)=0$. Then

$$0 = |W(\lambda_o)|^2 = |W(\overline{f_-(\lambda_o)},f_+(\lambda_o))|^2 \tag{2.15}$$

implies the contradiction $f_+(\lambda_o,x)\equiv 0$ because of Eqs. (2.9) and (2.10). Finally W cannot vanish on $\mathbb{C}\backslash\mathbb{R}$ since by the estimates (2.4) H would have a nonreal eigenvalue.

Lemma 2.2. Assume H(1) and suppose that $d_-(\lambda_o)=0$ for some $\lambda_o\in\mathbb{C}\backslash\{V_-\}$. Then $\lambda_o\in(-\infty,V_-)$ and

$$\dot{d}_-(\lambda_o) = [2ik_-(\lambda_o)]^{-1}\int_{\mathbb{R}} dx f_-(\lambda_o,x)f_+(\lambda_o,x) \neq 0. \tag{2.16}$$

Proof. We have

$$2ik_-(\lambda_o)\dot{d}_-(\lambda_o) = W(\dot{f}_-(\lambda_o),f_+(\lambda_o))_x + W(f_-(\lambda_o),\dot{f}_+(\lambda_o))_x, \quad x\in\mathbb{R} \tag{2.17}$$

and (differentiating Schrödinger's equation w.r. to z)

$$(\partial/\partial x)W(\dot{f}_-(z),f_+(z))_x = f_-(z,x)f_+(z,x),$$
$$(\partial/\partial x)W(f_-(z),\dot{f}_+(z))_x = -f_-(z,x)f_+(z,x), \quad x\in\mathbb{R},\ z\in\mathbb{C}. \tag{2.18}$$

Due to estimate (2.5)

$$\lim_{x\to-\infty} W(\dot{f}_-(\lambda_o),f_+(\lambda_o))_x = \lim_{x\to+\infty} W(f_-(\lambda_o),\dot{f}_+(\lambda_o))_x = 0 \tag{2.19}$$

and we arrive at

$$\int_{-\infty}^{0} dx(\partial/\partial x)W(\dot{f}_-(\lambda_0),f_+(\lambda_0))_x=\int_{-\infty}^{0} dxf_-(\lambda_0,x)f_+(\lambda_0,x),$$

$$-\int_0^{\infty} dx(\partial/\partial x)W(f_-(\lambda_0),\dot{f}_+(\lambda_0))_x=\int_0^{\infty} dxf_-(\lambda_0,x)f_+(\lambda_0,x). \tag{2.20}$$

This implies

$$\dot{W}(\lambda_0)=2ik_-(\lambda_0)\dot{d}_-(\lambda_0)=\mu\int_{\mathbb{R}} dx\,[f_+(\lambda_0,x)]^2\neq 0 \tag{2.21}$$

since by assumption $f_-(\lambda_0,x)=\mu f_+(\lambda_0,x)$, $\mu\neq 0$ and $f_+(\lambda_0,x)$ is real-valued $(\lambda_0<V_-\leq V_+)$.

Lemma 2.3. Assume H(0). Then

(a) $d_-(z)=1+O(|z|^{-1/2})$ as $|z|\to\infty$, $z\in\mathbb{C}$.

(b) $c_-(\lambda)=O(\lambda^{-1/2})$ as $\lambda\to+\infty$.

Proof. We indicate the proof of (a).

$$2ik_-(z)d_-(z)=W(f_-(z),f_+(z))_{x=0}$$

$$=\{1+\int_0^{-\infty} dx'k_-^{-1}\sin(k_-x')[V(x')-V_-]f_-(z,x')\}\{ik_+-\int_0^{\infty} dx'\cos(k_+x')[V(x')-V_+]f_+(z,x')\}$$

$$-\{1+\int_0^{\infty} dx'k_+^{-1}\sin(k_+x')[V(x')-V_+]f_+(z,x')\}\{-ik_--\int_0^{-\infty} dx'\cos(k_-x')[V(x')-V_-]f_-(z,x')\}$$

$$=i(k_++k_-)+O(1) \text{ as } |z|\to\infty \tag{2.22}$$

where Eq. (2.2), its derivative w.r. to x and (cf. Eq. (2.4))

$$|f_\pm(z,x)|\leq c_\pm e^{\mp(\mathrm{Im}k_\pm)x},\ x\gtrless 0,\ z\in\mathbb{C}\backslash\{V_\pm\} \tag{2.23}$$

with $c_\pm$ independent of z has been used. Since

$$k_-^{-1}(z)k_+(z)=1+O(z^{-1}) \text{ as } |z|\to\infty \tag{2.24}$$

we get (a). Assertion (b) follows similarly.

Concerning the threshold behavior we state

Lemma 2.4. Assume H(1). Then $d_-(z)$ has at most a finite number of zeros in $\mathbb{C}$ if either

(a) $W(V_-)\neq 0$

or

(b) $W(V_-)=0$ and $\lim\limits_{\substack{z\to V_- \\ \mathrm{Im}(z-V_-)^{1/2}\geq 0}} (z-V_-)^{-1/2}W(z)$ exists and is nonzero.

Proof. By Lemma 2.3 there exists an $R_o>0$ such that $|d_-(z)|\geq 1/2$ for $|z|\geq R_o$. By Lemma 2.1 the zeros of d_- lie in $(-\infty,V_-]$. If d_- has infinitely many zeros in $(-\infty,V_-]$, then due to analyticity of d_- in $\mathbb{C}\setminus[V_-,\infty)$, V_- must be an accumulation point of zeros of d_- which contradicts (a) or (b).

Lemma 2.5. Assume H(2). Then either

$W(V_-)\neq 0$

or

$$W(V_-)=0 \text{ and } \lim_{\substack{z\to V_- \\ \mathrm{Im}(z-V_-)^{1/2}\geq 0}} (z-V_-)^{-1/2}W(z)=i\gamma,\ \gamma\in\mathbb{R}\setminus\{0\}.$$

Proof. We only treat the case $V_-<V_+$ since the case $V_-=V_+$ is well known (cf. e.g. [18]). Then

$$k_+(z)=i|V_+-V_-|^{1/2}-i(z-V_-)|V_+-V_-|^{-1/2}/2+O((z-V_-)^2) \tag{2.25}$$

and hence the estimates (2.6)-(2.8), the first part in Eq. (2.22) and

$$\begin{aligned} &f_-(z,x)=f_-(V_-,x)+k_-((\partial/\partial k_-)f_-)(\tilde{z},x), \\ &f_+(z,x)=f_+(V_-,x)+k_-^2((\partial^2/\partial k_-^2)f_+)(\tilde{\tilde{z}},x)/2, \\ &\qquad |z-V_-| \text{ small enough} \end{aligned} \tag{2.26}$$

implies

$$\begin{aligned} W(z)&=W(f_-(z),f_+(z))_{x=0}=W(V_-) \\ &+i\{1+\int_0^\infty dx'|V_+-V_-|^{-1/2}\sinh(|V_+-V_-|^{1/2}x')[V(x')-V_+]f_+(V_-,x')\}(z-V_-)^{1/2} \\ &+o((z-V_-)^{1/2}) \\ &\equiv i\gamma(z-V_-)^{1/2}+o((z-V_-)^{1/2}),\ \gamma\in\mathbb{R}. \end{aligned} \tag{2.27}$$

Thus $d_-(V_-)=\gamma/2$. For $\lambda\in(V_-,V_+)$ we have

$c_-(\lambda)=\overline{d_-(\lambda)}$ in Eq. (2.9) and hence

$$f_+(\lambda,x)=2\mathrm{Re}[d_-(\lambda)\overline{f_-(\lambda,x)}],\ \lambda\in(V_-,V_+),\ x\in\mathbb{R}. \tag{2.28}$$

As $\lambda\downarrow V_-$ we obtain

$$f_+(V_-,x)=\gamma f_-(V_-,x),\ x\in\mathbb{R} \tag{2.29}$$

(since all quantities in Eq. (2.28) become real as $\lambda\downarrow V_-$) and hence $\gamma\neq 0$.

Remark 2.6. The results of Lemmas 2.1-2.5 (occasionally under slightly stronger assumptions) have been proven in [15] on the basis of integral equations different from Eq. (2.2) by following the case $V_\pm=0$ treated in [18]. In fact the approach chosen in [15] and [18] is most suited in the context of the inverse scattering problem which is their main concern. For completeness we sketched the corresponding proofs based on Eq. (2.2).

In the special case $V_\pm=0$, $k_\pm=k$ one obtains ($x\in\mathbb{R}$ arbitrary)

$$W(f_-(z),f_+(z)) = 2ik - \int_{-\infty}^{x} dx' e^{ikx'} V(x') f_-(z,x') - \int_{x}^{\infty} dx' e^{-ikx'} V(x') f_+(z,x')$$

$$-(2ik)^{-1} \int_{-\infty}^{x} dx' e^{-ikx'} V(x') f_-(z,x') \int_{x}^{\infty} dx'' e^{ikx''} V(x'') f_+(z,x''), \quad z\in\mathbb{C}\setminus\{0\} \tag{2.30}$$

Taking $x\to\pm\infty$ we recover the familiar expressions ([14])

$$W(f_-(z),f_+(z)) = 2ik - \int_{\mathbb{R}} dx' e^{\pm ikx'} V(x') f_\mp(z,x'), \quad z\in\mathbb{C}\setminus\{0\}. \tag{2.31}$$

Concerning spectral properties of H we state ($\sigma(.)$, $\sigma_{ess}(.)$, $\sigma_{ac}(.)$, $\sigma_{sc}(.)$, $\sigma_p(.)$ denote the spectrum, essential spectrum, absolutely continuous spectrum, singularly continuous spectrum and point spectrum respectively)

<u>Theorem 2.7.</u> (a) Assume H(0). Then H is bounded from below and

$$\sigma_{ess}(H)=\sigma_{ac}(H)=[V_-,\infty), \quad \sigma_{sc}(H)=\emptyset \tag{2.32}$$

(since we choose $V_-\leq V_+$). Moreover H has simple spectrum in (V_-,V_+) (in case $V_-<V_+$) and spectral multiplicity two in (V_+,∞). In addition all bound states $\lambda_o\in(-\infty,V_-)$ of H are determined by

$$W(f_-(\lambda_o),f_+(\lambda_o))=0 \tag{2.33}$$

and

$$\sigma_p(H)\cap(V_-,\infty)=\emptyset. \tag{2.34}$$

(b) Assume H(1). Then $\sigma_p(H)$ contains only simple eigenvalues in $(-\infty,V_-)$.
(c) Assume H(2). Then H has finitely many bound states and no threshold eigenvalue i.e.

$$V_-\notin\sigma_p(H). \tag{2.35}$$

If $W(f_-(V_-),f_+(V_-))=0$ then H has a threshold resonance whose wave function (after suitable normalization) $\psi(V_-,x)$ behaves like

$$|\psi(V_-,x)-1|\leq c(1+|x|)^{-1}, \quad x\leq 0,$$
$$|\psi(V_-,x)|\leq c' e^{-|\mathrm{Im}\,k_+(V_-)|x}, \quad x\geq 0. \tag{2.36}$$

<u>Proof.</u> The result (2.32) is due to [49]. Since $f_+(\lambda,x)$ is real and

decays exponentially as $x\to+\infty$ for $\lambda\in(V_-,V_+)$, H has simple spectrum in (V_-,V_+). By Eq. (2.2) the spectral multiplicity of H equals two in (V_+,∞). Eq. (2.34) follows from Lemma 2.1. Simplicity of the bound states follows from Lemma 2.2. Since $f_-(V_-,x)$ stays bounded as $x\to-\infty$ (cf. Eq. (2.7)) the second linearly independent solution is easily shown to grow linearly as $x\to-\infty$ using H(2). Thus Eq. (2.35) holds. Eqs. (2.36) are clear from Eqs. (2.6) and (2.7).

For spectral properties of H in the presence of dilation analytic interactions see [12].

Assuming H(0), stationary scattering theory for H now becomes extremely simple. We start with the energy regime where H has spectral multiplicity two.

a) $\lambda>V_+$.

Then the on-shell scattering matrix $S(\lambda)$ in $\mathbb{C}^2$ reads

$$S(\lambda)=\begin{bmatrix} T^{\ell}(\lambda) & R^{r}(\lambda) \\ R^{\ell}(\lambda) & T^{r}(\lambda)\end{bmatrix}, \quad \lambda>V_+ \tag{2.37}$$

where the transmission and reflection coefficients from the left and right are given by

$$T^{\ell}(\lambda)=T^{r}(\lambda)\equiv T(\lambda)=\frac{(k_-/k_+)^{1/2}}{d_+(\lambda)}=\frac{(k_+/k_-)^{1/2}}{d_-(\lambda)}=\frac{2i(k_+k_-)^{1/2}}{W(f_-(\lambda),f_+(\lambda))}, \tag{2.38}$$

$$R^{\ell}(\lambda)=c_-(\lambda)/d_-(\lambda)=-W(\overline{f_-(\lambda)},f_+(\lambda))/W(f_-(\lambda),f_+(\lambda)),$$
$$R^{r}(\lambda)=c_+(\lambda)/d_+(\lambda)=-W(f_-(\lambda),\overline{f_+(\lambda)})/W(f_-(\lambda),f_+(\lambda));\ \lambda>V_+. \tag{2.39}$$

Unitarity of $S(\lambda)$, $\lambda>V_+$ now trivially follows from identity (2.12) and Eq. (2.14). In particular

$$|T(\lambda)|^2+|R^{\ell}(\lambda)|^2=1=|T(\lambda)|^2+|R^{r}(\lambda)|^2,$$
$$|R^{\ell}(\lambda)|=|R^{r}(\lambda)|;\ \lambda>V_+. \tag{2.40}$$

The phase shift $\delta(\lambda)$, $\lambda>V_+$ is given by

$$e^{2i\delta(\lambda)}=\det S(\lambda)=T(\lambda)/\overline{T(\lambda)},\ \lambda>V_+. \tag{2.41}$$

In the energy regime where H has simple essential spectrum we have the following results (assuming $V_-<V_+$).

b) $V_-<\lambda<V_+$.

Since $f_+(\lambda,x)$, $x\in\mathbb{R}$ is real for $\lambda\in(V_-,V_+)$ we infer

$$S(\lambda)=e^{2i\delta(\lambda)}=\frac{\overline{d_-(\lambda)}}{d_-(\lambda)}=-\frac{\overline{W(f_-(\lambda),f_+(\lambda))}}{W(f_-(\lambda),f_+(\lambda))}, \quad V_-<\lambda<V_+. \tag{2.42}$$

In particular

$$R^{\ell}(\lambda)=c_-(\lambda)/d_-(\lambda)=S(\lambda), \quad V_-<\lambda<V_+ \tag{2.43}$$

and hence

$$|R^{\ell}(\lambda)|=|S(\lambda)|=1, \; V_-<\lambda<V_+ \tag{2.44}$$

proves total reflection from the left for $\lambda\in(V_-,V_+)$.

As mentioned in the introduction the result (2.44) has been derived in [16] (cf. also [6]) by different means. In the above approach total reflection for energies $\lambda\in(V_-,V_+)$, i.e. for energies where H has spectral multiplicity one, is a simple consequence of the fact that the Jost soution $f_+(\lambda,x)$ is real-valued and unbounded as $x\to-\infty$.

In both cases a) and b) we have

$$e^{2i\delta(\lambda)}=\det S(\lambda)=\frac{\overline{d_-(\lambda)}}{d_-(\lambda)}=-\frac{\overline{W(f_-(\lambda),f_+(\lambda))}}{W(f_-(\lambda),f_+(\lambda))}, \quad \lambda>V_-. \tag{2.45}$$

Finally we discuss asymptotically similar potentials V_1 and V_2 and describe the corresponding relative scattering formalism. We introduce hypothesis

<u>$H_{12}(\alpha,\beta)$.</u> Assume that V_j, j=1,2 fulfill $H(\alpha)$ for some $\alpha\geq 0$ and fixed $V_\pm\in\mathbb{R}$. Moreover suppose that

$$\int_{\mathbb{R}} dx(1+|x|^{\beta})|V_{12}(x)|<\infty, \quad V_{12}(x)=V_1(x)-V_2(x)$$

for some $\beta\geq 0$.

Given $H_{12}(0,0)$ (which will always be assumed for the rest of this section) we introduce in $L^2(\mathbb{R})$ the form sums

$$H_j=-\frac{d^2}{dx^2}\dotplus V_j, \; j=1,2. \tag{2.46}$$

Then the Jost solutions $f_{j\pm}$ associated with H_j, j=1,2 (in addition to Eq. (2.2)) uniquely fulfill Volterra - and Fredholm integral equations of the type

$$f_{1\pm}(z,x)=f_{2\pm}(z,x)-\int_x^{\pm\infty} dx' g_{2-+}(z,x,x')V_{12}(x')f_{1\pm}(z,x'),$$
$$z\in\mathbb{C}\backslash\{\sigma_p(H_2)\cup\{V_-\}\} \tag{2.47}$$

where

$$g_{2-+}(z,x,x')=[f_{2+}(z,x)f_{2-}(z,x')-f_{2-}(z,x)f_{2+}(z,x')]/W(f_{2-}(z),f_{2+}(z)),$$
$$z\in\mathbb{C}\backslash\{\sigma_p(H_2)\cup\{V_-\}\} \tag{2.48}$$

and

$$f_{1\pm}(z,x)= [T_{12}(z)]^{-1}f_{2\pm}(z,x)-\int_{\mathbb{R}}dx'g_2(z,x,x')V_{12}(x')f_{1\pm}(z,x'),$$
$$z\in\mathbb{C}\backslash\{\sigma_p(H_2)\cup\{V_-\}\} \tag{2.49}$$

where

$$g_2(z,x,x') = \frac{-1}{W(f_{2-}(z),f_{2+}(z))}\begin{cases} f_2(z,x)f_{2-}(z,x'), & x\geq x' \\ f_{2+}(z,x')f_{2-}(z,x), & x\leq x' \end{cases},$$
$$z\in\mathbb{C}\backslash\{\sigma_p(H_2)\cup\{V_-\}\}, \tag{2.50}$$

$$g_2(z)=(H_2-z)^{-1},\quad z\in\rho(H_2) \tag{2.51}$$

($\rho(.)$ the resolvent set) and

$$T_{12}(z)=T_1(z)/T_2(z)=W(f_{2-}(z),f_{2+}(z))/W(f_{1-}(z),f_{1+}(z)),$$
$$z\in\mathbb{C}\backslash\{\sigma_p(H_1)\cup\{V_-\}\}. \tag{2.52}$$

Assuming $H_{12}(0,0)$, the asymptotic behavior of $f_{1\pm}(z,x)$ as $x\to\pm\infty$ yields

$$d_{1\pm}(z)=d_{2\pm}(z)-(2ik_\pm)^{-1}\int_{\mathbb{R}}dxf_{2\pm}(z,x)V_{12}(x)f_{1\mp}(z,x), \tag{2.53}$$

$$c_{1\pm}(z)=c_{2\pm}(z)+(2ik_\pm)^{-1}\int_{\mathbb{R}}dx\overline{f_{2\pm}(z,x)}V_{12}(x)f_{1\mp}(z,x);\quad z\in\mathbb{C}\backslash\{V_\pm\}. \tag{2.54}$$

Thus

$$\begin{aligned}T_{12}(z)&=d_{2-}(z)/d_{1-}(z)\\ &=\{1-[2ik_-d_{2-}(z)]^{-1}\int_{\mathbb{R}}dxf_{2-}(z,x)V_{12}(x)f_{1+}(z,x)\}^{-1},\quad z\in\mathbb{C}\backslash\{\sigma_p(H_1)\cup\{V_-\}\}.\end{aligned} \tag{2.55}$$

Similarly

$$e^{2i\delta_1(\lambda)}=\det S_1(\lambda)=\overline{d_{1-}(\lambda)}/d_{1-}(\lambda)$$
$$=\frac{\overline{d_{2-}(\lambda)}}{d_{2-}(\lambda)}\,\frac{1+[2ik_-\overline{d_{2-}(z)}]^{-1}\int_{\mathbb{R}}dx\overline{f_{2-}(z,x)}V_{12}(x)\overline{f_{1+}(z,x)}}{1-[2ik_-d_{2-}(z)]^{-1}\int_{\mathbb{R}}dxf_{2-}(z,x)V_{12}(x)f_{1+}(z,x)}=e^{2i\delta_2(\lambda)}e^{2i\delta_{12}(\lambda)},$$
$$\lambda>V_- \tag{2.56}$$

where

$$e^{2i\delta_{12}(\lambda)}=T_{12}(\lambda)/\overline{T_{12}(\lambda)}\ ,\quad \lambda>V_- \tag{2.57}$$

describes the relative phase shift for the pair (H_1,H_2).

Finally we connect $T_{12}(z)$ and the Fredholm determinant associated with (H_1,H_2).

Theorem 2.8. Assume $H_{12}(1,0)$ and define

$$v_{12}(x)=|V_{12}(x)|^{1/2},\quad u_{12}(x)=v_{12}(x)\,\mathrm{sgn}(V_{12}(x)). \tag{2.58}$$

Then

$$\begin{aligned}[T_{12}(z)]^{-1}&=W(f_{1-}(z),f_{1+}(z))/W(f_{2-}(z),f_{2+}(z))\\ &=\det[1+u_{12}(H_2-z)^{-1}v_{12}],\quad z\in\rho(H_2).\end{aligned} \tag{2.59}$$

Proof. Hypothesis $H_{12}(0,0)$ implies that

$$(H_1-z)^{-1}-(H_2-z)^{-1}=-(H_2-z)^{-1}v_{12}[1+u_{12}(H_2-z)^{-1}v_{12}]^{-1}u_{12}(H_2-z)^{-1}\in\mathcal{B}_1(L^2(\mathbb{R})),$$
$$z\in\rho(H_1)\cap\rho(H_2) \tag{2.60}$$

(Here $\mathcal{B}_p(.)$ denote the usual trace ideals [45]). This is seen as follows. First of all Eq. (2.2) and H(0) imply

$$|f_\pm(\lambda,x)|\geq e^{\kappa_\pm x}/2,\quad x\gtrless x_0(\lambda)\gtrless 0,\quad \lambda<V_-,\quad \kappa_\pm=|V_\pm-\lambda|^{1/2} \tag{2.61}$$

for $|x_0(\lambda)|$ large enough. Introducing

$$\tilde{f}_\pm(\lambda,x)=f_\pm(\lambda,x)\int_{x_0(\lambda)}^{x}dx'\,[f_\pm(\lambda,x')]^{-2},\quad x\gtrless x_0(\lambda)\gtrless 0,\quad \lambda<V_- \tag{2.62}$$

then

$$W(f_\pm(\lambda),\tilde{f}_\pm(\lambda))=1,\quad \lambda<V_- \tag{2.63}$$

and hence

$$|f_\pm(\lambda,x)|\leq c_\pm(\lambda)e^{\mp\kappa_\mp x},\quad x\lesseqgtr 0,\quad \lambda<V_-. \tag{2.64}$$

Consequently

$$|g_2(\lambda,x,x')|\leq c_2(\lambda)e^{-\kappa_-|x-x'|},\quad \lambda<V_-. \tag{2.65}$$

The estimate (2.65) in terms of the "free" resolvent proves

$$u_{12}(H_2-z)^{-1},\ u_{12}(H_2-z)^{-1}v_{12}\in\mathcal{B}_2(L^2(\mathbb{R})),\quad z\in\rho(H_2) \tag{2.66}$$

and hence the result (2.60). Actually, using

$$g_2(\lambda,x,x')>0 \text{ for } \lambda<\inf\sigma(H_2) \tag{2.67}$$

one can follow [42], p.65 and prove

$$u_{12}(H_2-z)^{-1}v_{12}\in\mathcal{B}_1(L^2(\mathbb{R})),\ z\in\rho(H_2). \tag{2.68}$$

From now on are can follow the arguments of [36] (in the special case $V_1=V$, $V_2=0$) step by step using Eqs. (2.47) and (2.49). This yields Eq. (2.59) for $\lambda<\inf\{\sigma(H_1)\cup\sigma(H_2)\}$. Analytic continuation of both sides (using H(1,0)) w.r. to λ then completes the proof.

Remark 2.9. Obviously the left hand side of Eq. (2.59) has a continuous extension to $\lambda\in\mathbb{R}\setminus\sigma_p(H_2)$. But since $g_2(\lambda,x,x')$ in general will not be of convolution type one cannot simply follow Proposition 5.6 of [45] to extend the validity of Eq. (2.59) on the cut (V_-,∞).

3. The case $V(x)\to+\infty$ as $x\to+\infty$

We start with a basic result (for a proof cf. e.g. [41]).

Lemma 3.1. Let $W\in L^1_{loc}(\mathbb{R})$ be real-valued, $W\geq 1$ a.e. on $[x_o,\infty)$ for some $x_o\in\mathbb{R}$. Then the equation

$$-\psi''(x)+W(x)\psi(x)=0$$

has two solutions $f,\tilde{f}$ with the properties

(a) $f,f',\tilde{f},\tilde{f}'\in AC_{loc}(\mathbb{R})$, $f,\tilde{f}>0$ on $[x_o,\infty)$.

(b) $f'\leq -f$, $\tilde{f}'\geq\tilde{f}$ on $[x_o,\infty)$

(i.e. $f(x)\leq f(x_o)e^{-(x-x_o)}$, $\tilde{f}(x)\geq\tilde{f}(x_o)e^{(x-x_o)}$ on $[x_o,\infty)$).

(c) $f(x)\tilde{f}(x)\leq C$ on $[x_o,\infty)$.

(Here $AC_{loc}(\Omega)$ denotes the set of locally absolutely continuous functions on $\Omega\subseteq\mathbb{R}$). Next we introduce hypothesis

$H_-(\alpha)$. Let V be a real-valued function on $\mathbb{R}$, $V\in L^1_{loc}(\mathbb{R})$, $V_-\in\mathbb{R}$ and

$$\int_{-\infty}^{0} dx(1+|x|^\alpha)|V(x)-V_-|<\infty$$

for some $\alpha\geq 0$. Moreover assume that $\lim_{x\to+\infty}V(x)=+\infty$.

Given $H_-(0)$ (which will always be assumed in the following) we define $H=-\frac{d^2}{dx^2}+V$, $f_-(z,x)$ and $k_-(z)$, $z\in\mathbb{C}$ as in Eqs. (2.1)-(2.3). Again $f_-(z,x)$ is analytic in $z\in\mathbb{C}\setminus[V_-,\infty)$ using $H_-(1)$. Next fix $\lambda>V_-$ and let $f_+(\lambda,x)$, $\tilde{f}_+(\lambda,x)$ denote the solutions of

$$-\chi''(\lambda,x)+[V(x)-\lambda]\chi(\lambda,x)=0, \quad \lambda>V_- \tag{3.1}$$

according to Lemma 3.1 with some $x_0(\lambda)\in\mathbb{R}$. Then

$$f_+(\lambda,x)=c_-(\lambda)f_-(\lambda,x)+d_-(\lambda)\overline{f_-(\lambda,x)}, \quad \lambda>V_-, \tag{3.2}$$

$$2ik_-d_-(\lambda)=W(f_-(\lambda),f_+(\lambda)), \quad \lambda>V_- \tag{3.3}$$

and

$$c_-(\lambda)=\overline{d_-(\lambda)}, \quad \lambda>V_- \tag{3.4}$$

since $f_+(\lambda,x)$ is real. Now one can follow Sect. 2 step by step. E.g. we get

<u>Theorem 3.2.</u> (a) Assume $H_-(0)$. Then H is bounded from below and

$$\sigma_{ess}(H)=\sigma_{ac}(H)=[V_-,\infty), \quad \sigma_{sc}(H)=\emptyset. \tag{3.5}$$

Moreover H has simple spectrum in (V_-,∞) and

$$\sigma_p(H)\cap(V_-,\infty)=\emptyset. \tag{3.6}$$

All eigenvalues $\lambda_0\in(-\infty,V_-)$ of H are determined by

$$W(f_-(\lambda_0),f_+(\lambda_0))=0. \tag{3.7}$$

(b) Assume $H_-(1)$. Then $\sigma_p(H)$ contains only simple bound states in $(-\infty,V_-)$.

Due to the simplicity of $\sigma(H)\cap(V_-,\infty)$ the on-shell scattering matrix in analogy to case b) of Sect. 2 becomes

$$S(\lambda)=e^{2i\delta(\lambda)}=\frac{\overline{d_-(\lambda)}}{d_-(\lambda)}=-\frac{\overline{W(f_-(\lambda),f_+(\lambda))}}{W(f_-(\lambda),f_+(\lambda))}, \quad \lambda>V_-. \tag{3.8}$$

In particular

$$R^\ell(\lambda)=c_-(\lambda)/d_-(\lambda)=S(\lambda),$$

$$|R^\ell(\lambda)|=|S(\lambda)|=1; \quad \lambda>V_- \tag{3.9}$$

proves again total reflection from the left for all scattering energies $\lambda>V_-$ (i.e. the incoming particle from the left will bounce off the wall no matter what energy it carries). This result has been obtained in [16]. In the above approach it is a simple consequence of the fact that $f_+(\lambda,x)$ is real-valued and unbounded as $x\to-\infty$ for all $\lambda>V_-$ or equivalently that H has simple spectrum in (V_-,∞).

Since asymptotically similar potentials can be discussed as in Sect. 2 we omit the details.

4. Half crystals

We first review some results from Floquet theory (cf. e.g. [10], [19], [31], [38], [43] and [48]). Assume

HC: Let W be a real-valued periodic function on $\mathbb{R}$,

$W(x+a)=W(x)$, $x\in\mathbb{R}$ for some $a>0$ and assume $W\in L^1_{loc}(\mathbb{R})$.

Given HC, the Hamiltonian H_a in $L^2(\mathbb{R})$ defined by

$$H_a=-\frac{d^2}{dx^2}\dot{+}W \tag{4.1}$$

admits Floquet solutions $g_\pm(z,x)$ with the following properties:

$$\begin{aligned}
&g_\pm(z),\ g_\pm'(z)\in AC_{loc}(\mathbb{R}),\\
&-g_\pm''(z,x)+[W(x)-z]g_\pm(z,x)=0,\ z\in\mathbb{C},\ x\in\mathbb{R},\\
&g_\pm(z,x)=e^{\pm i\theta(z)x}p_\pm(z,x),\ p_\pm(z,x+a)=p_\pm(z,x),\\
&\lambda\in\sigma(H_a)\ \text{iff}\ \theta(\lambda)\in\mathbb{R},\\
&\lambda\in\mathbb{R}\backslash\sigma(H_a)\ \text{iff}\ \theta(\lambda)\in i(0,\infty),\\
&g_+(\lambda,x)=\overline{g_-(\lambda,x)},\ \lambda\in\sigma(H_a),\\
&g_\pm(\lambda,x)\ \text{are real-valued for}\ \lambda\in\mathbb{R}\backslash\overset{\circ}{\sigma}(H_a)
\end{aligned} \tag{4.2}$$

($\overset{\circ}{\Omega}$ the interior of a set $\Omega\subset\mathbb{R}$).

Next introduce

$$\psi(z,x)=k^{-1}\sin kx+\int_0^x dx'k^{-1}\sin[k(x-x')]W(x')\psi(z,x'),\ 0\leq x\leq a,$$

$$\psi(z,0)=0,\ \psi'(z,0)=1,\ z\in\mathbb{C} \tag{4.3}$$

where

$$k(z)=z^{1/2},\ \operatorname{Im}k(z)\geq 0. \tag{4.4}$$

One can show that

$$\psi(\lambda,a)\neq 0\ \text{for}\ \lambda\in\overset{\circ}{\sigma}(H_a) \tag{4.5}$$

and that ([38])

$$W(g_+(\lambda),\overline{g_+(\lambda)})=\begin{cases}-2i\psi(\lambda,a)\sin[\theta(\lambda)a], & \lambda\in\overset{\circ}{\sigma}(H_a)\\ 0, & \lambda\in\mathbb{R}\backslash\overset{\circ}{\sigma}(H_a).\end{cases} \tag{4.6}$$

Finally the spectrum of H_a is given by ([43])

$$\sigma(H_a)=\sigma_{ac}(H_a)=\bigcup_{n=1}^{\infty}[\alpha_n,\beta_n],\ \sigma_p(H_a)=\sigma_{sc}(H_a)=\emptyset \tag{4.7}$$

where

$$\alpha_n = \begin{cases} E_n(0), & n \text{ odd} \\ E_n(\pi/a), & n \text{ even} \end{cases}, \quad \beta_n = \begin{cases} E_n(\pi/a), & n \text{ odd} \\ E_n(0), & n \text{ even} \end{cases}, \quad n \in \mathbb{N} \tag{4.8}$$

and $E_n(\theta)$, $0 \leq \theta \leq \pi/a$, $n \in \mathbb{N}$ are the eigenvalues of $H(\theta)$ in $L^2(0,a)$ ordered in magnitude

$$H(\theta) = -\frac{d^2}{dx^2}\Big|_\theta \dot{+} W,$$

$$\mathcal{D}\Big(-\frac{d^2}{dx^2}\Big|_\theta\Big) = \{f \in L^2(0,a) \,|\, f, f' \in AC_{loc}(0,a);\ f(a_-) = e^{i\theta a} f(0_+),$$

$$f'(a_-) = e^{i\theta a} f'(0_+);\ f', f'' \in L^2(0,a)\},\ 0 \leq \theta \leq \pi/a. \tag{4.9}$$

Now we are in position to introduce the system composed of two half crystals. Let W_j fulfill HC with periods $a_j > 0$, $j=1,2$ and define in $L^2(\mathbb{R})$

$$H = -\frac{d^2}{dx^2} \dot{+} V, \qquad V(x) = \begin{cases} W_1(x), & x<0 \\ W_2(x), & x>0. \end{cases} \tag{4.10}$$

Then

$$\sigma_{ess}(H) = \sigma_{ac}(H) = \sigma(H_{a_1}) \cup \sigma(H_{a_2}),\ \sigma_{sc}(H) = \emptyset \tag{4.11}$$

and H has spectral multiplicity two in $\{\sigma(H_{a_1}) \cap \sigma(H_{a_2})\}^o$ and simple spectrum in $\{\sigma_{ac}(H) \setminus \{\sigma(H_{a_1}) \cap \sigma(H_{a_2})\}\}^o$. These results easily follow by considering Jost solutions $f_\pm(z,x)$ associated with H. In fact we get in the case of spectral multiplicity two

a) $\lambda \in \{\sigma(H_{a_1}) \cap \sigma(H_{a_2})\}^o$.

$$f_-(\lambda,x) = \begin{cases} g_{1-}(\lambda,x), & x<0 \\ c_+(\lambda) g_{2+}(\lambda,x) + d_+(\lambda)\overline{g_{2+}(\lambda,x)}, & x>0, \end{cases}$$

$$f_+(\lambda,x) = \begin{cases} c_-(\lambda) g_{1-}(\lambda,x) + d_-(\lambda)\overline{g_{1-}(\lambda,x)}, & x<0 \\ g_{2+}(\lambda,x), & x>0 \end{cases} \tag{4.12}$$

implying

$$d_-(\lambda) = \frac{W(g_{2+}(\lambda), g_{1-}(\lambda))_{x=0}}{W(\overline{g_{1-}(\lambda)}, g_{1-}(\lambda))_{x=0}}, \quad c_-(\lambda) = \frac{W(g_{2+}(\lambda), \overline{g_{1-}(\lambda)})_{x=0}}{W(g_{1-}(\lambda), \overline{g_{1-}(\lambda)})_{x=0}},$$

$$d_+(\lambda) = \frac{W(g_{1-}(\lambda), g_{2+}(\lambda))_{x=0}}{W(\overline{g_{2+}(\lambda)}, g_{2+}(\lambda))_{x=0}}, \quad c_+(\lambda) = \frac{W(g_{1-}(\lambda), \overline{g_{2+}(\lambda)})_{x=0}}{W(g_{2+}(\lambda), \overline{g_{2+}(\lambda)})_{x=0}};$$

$$\lambda \in \{\sigma(H_{a_1}) \cap \sigma(H_{a_2})\}^{\circ} \tag{4.13}$$

in order to guarantee $f_{\pm}(\lambda)$, $f'_{\pm} \in AC_{loc}(\mathbb{R})$. In this case the on-shell scattering matrix $S(\lambda)$ in $\mathbb{C}^2$ reads

$$S(\lambda) = \begin{bmatrix} T^{\ell}(\lambda) & R^{r}(\lambda) \\ R^{\ell}(\lambda) & T^{r}(\lambda) \end{bmatrix}, \quad \lambda \in \{\sigma(H_{a_1}) \cap \sigma(H_{a_2})\}^{\circ}, \tag{4.14}$$

$$T^{\ell}(\lambda) = T^{r}(\lambda) \equiv T(\lambda) = \frac{\{-\psi_1(\lambda)\sin[\theta_1(\lambda)a_1]/\psi_2(\lambda)\sin[\theta_2(\lambda)a_2]\}^{1/2}}{d_+(\lambda)}$$

$$= \frac{\{-\psi_2(\lambda)\sin[\theta_2(\lambda)a_2]/\psi_1(\lambda)\sin[\theta_1(\lambda)a_1]\}^{1/2}}{d_-(\lambda)}$$

$$= \frac{2i\{-\psi_1(\lambda)\psi_2(\lambda)\sin[\theta_1(\lambda)a_1]\sin[\theta_2(\lambda)a_2]\}^{1/2}}{W(g_{1-}(\lambda),g_{2+}(\lambda))_{x=0}}, \tag{4.15}$$

$$R^{\ell}(\lambda) = \frac{c_-(\lambda)}{d_-(\lambda)} = -\frac{W(\overline{g_{1-}(\lambda)},g_{2+}(\lambda))_{x=0}}{W(g_{1-}(\lambda),g_{2+}(\lambda))_{x=0}},$$

$$R^{r}(\lambda) = \frac{c_+(\lambda)}{d_+(\lambda)} = -\frac{W(g_{1-}(\lambda),\overline{g_{2+}(\lambda)})_{x=0}}{W(g_{1-}(\lambda),g_{2+}(\lambda))_{x=0}}; \quad \lambda \in \{\sigma(H_{a_1}) \cap \sigma(H_{a_2})\}^{\circ}. \tag{4.16}$$

Unitarity of $S(\lambda)$ again follows immediately from identity (2.12) and Eq. (4.6). The corresponding phase shift is given by

$$e^{2i\delta(\lambda)} = \det S(\lambda) = \frac{T(\lambda)}{\overline{T(\lambda)}} = -\frac{\overline{W(g_{1-}(\lambda),g_{2+}(\lambda))_{x=0}}}{W(g_{1-}(\lambda),g_{2+}(\lambda))_{x=0}},$$

$$\lambda \in \{\sigma(H_{a_1}) \cap \sigma(H_{a_2})\}^{\circ}. \tag{4.17}$$

Now we turn to the cases where H has simple essential spectrum.

b) $\lambda \in \{\sigma_{ac}(H) \backslash \sigma(H_{a_2})\}^{\circ}$.

Here $f_+(\lambda,x)$ and $c_-(\lambda)$, $d_-(\lambda)$ are given as in Eqs. (4.12) and (4.13) and since $g_{2\pm}(\lambda,x)$ are real-valued we obtain

$$S(\lambda) = e^{2i\delta(\lambda)} = \frac{\overline{d_-(\lambda)}}{d_-(\lambda)} = -\frac{\overline{W(g_{1-}(\lambda),g_{2+}(\lambda))_{x=0}}}{W(g_{1-}(\lambda),g_{2+}(\lambda))_{x=0}}, \tag{4.18}$$

$$R^{\ell}(\lambda) = \frac{c_-(\lambda)}{d_-(\lambda)} = S(\lambda); \quad \lambda \in \{\sigma_{ac}(H) \backslash \sigma(H_{a_2})\}^{\circ}. \tag{4.19}$$

c) $\lambda \in \{\sigma_{ac}(H) \backslash \sigma(H_{a_1})\}^{\circ}$.

Here $f_-(\lambda,x)$ and $c_+(\lambda)$, $d_+(\lambda)$ are given as in Eqs. (4.12) and (4.13) and since now $g_{1\pm}(\lambda,x)$ is real-valued we infer

$$S(\lambda)=e^{2i\delta(\lambda)}=\frac{\overline{d_+(\lambda)}}{d_+(\lambda)}=-\frac{\overline{W(g_{1-}(\lambda),g_{2+}(\lambda))_{x=0}}}{W(g_{1-}(\lambda),g_{2+}(\lambda))_{x=0}}, \tag{4.20}$$

$$R^r(\lambda)=\frac{c_+(\lambda)}{d_+(\lambda)}=S(\lambda);\ \lambda\in\{\sigma_{ac}(H)\backslash\sigma(H_{a_1})\}^o. \tag{4.21}$$

In case b) (case c)) $R^\ell(\lambda)$ ($R^r(\lambda)$) is of modulus one implying total reflection from the left (right). In other words in case a) a particle coming from the left (right) penetrates with nonzero probability into the right (left) half crystal whereas in cases b) or c) the particle is totally reflected at the interface and hence is confined to the left or right half crystal respectively. For the original derivation of this result cf. again [16]. As in Sects. 2 and 3 total reflection in the case of spectral multiplicity one in our approach simply results from the fact that either $g_{1\pm}(\lambda,x)$ or $g_{2\pm}(\lambda,x)$ become real-valued and unbounded. In all cases a)-c) we have

$$e^{2i\delta(\lambda)}=\det S(\lambda)=-\frac{\overline{W(g_{1-}(\lambda),g_{2+}(\lambda))_{x=0}}}{W(g_{1-}(\lambda),g_{2+}(\lambda))_{x=0}},\ \lambda\in\overset{o}{\sigma}_{ac}(H). \tag{4.22}$$

Without going into details we note that spectral and scattering theory for the pair (H_1,H_2) in $L^2(\mathbb{R})$

$$H_1=-\frac{d^2}{dx^2}\dotplus W\dotplus V,\qquad H_2=-\frac{d^2}{dx^2}\dotplus W \tag{4.23}$$

where W fulfills HC and $(1+|\cdot|^\alpha)V\in L^1(\mathbb{R})$ (for suitable $\alpha\geq 0$) is real-valued, can be developed in complete analogy to the end of Sect. 2 ([20], [51]). In particular the connection between the relative transmission coefficient and the Fredholm determinant associated with H_1 and H_2 (cf. Theorem 2.8) extends to the present case ([38]).

5. Applications in supersymmetric quantum mechanics

On the basis of [5], [22] we first recall a few facts about supersymmetric quantum mechanics for one degree of freedom.

Let A be a closed, densely defined operator in a separable, complex Hilbert space $\mathcal{H}$ and define the pair

$$H_1=A^*A,\quad H_2=AA^* \tag{5.1}$$

of "bosonic" resp. "fermionic" Hamiltonians in $\mathcal{H}$. The corresponding supercharge Q and the supersymmetric Hamiltonian H in $\mathcal{H}\oplus\mathcal{H}$ then read

$$Q=\begin{bmatrix}0 & A^*\\ A & 0\end{bmatrix},\qquad H=Q^2=\begin{bmatrix}H_1 & 0\\ 0 & H_2\end{bmatrix}. \tag{5.2}$$

We shall use the following set of assumptions.

H(i). $(H_1-z_o)^{-1}-(H_2-z_o)^{-1} \in \mathcal{B}_1(\mathcal{H})$ for some $z_o \in \rho(H_1) \cap \rho(H_2)$.

H(ii). In addition to H(i) assume that $H_1=H_2\dot{+}V_{12}$ where V_{12} can be split into two parts $V_{12}=u_{12}v_{12}$ such that $u_{12}(H_2-z)^{-1}v_{12}$ is analytic w.r. to $z \in \rho(H_2)$ in $\mathcal{B}_1(\mathcal{H})$-norm and that $u_{12}(H_2-z_o)^{-1}$, $(H_2-z_o)^{-1}v_{12} \in \mathcal{B}_2(\mathcal{H})$ for some $z_o \in \rho(H_2)$.

H(iii). Assume H(ii) and

$$\lim_{\substack{|z|\to\infty \\ |\mathrm{Im} z|\neq 0}} \det[1+u_{12}(H_2-z)^{-1}v_{12}]=1.$$

Then we have (cf. [5])

Lemma 5.1. (a) Assume H(i). Then there exists a real-valued, measurable function ξ_{12} on $\mathbb{R}$ (Krein's spectral shift function [4], [33], [34]) unique up to a constant a.e. with

$(1+|.|^2)^{-1}\xi_{12} \in L^1(\mathbb{R})$ and

$$\mathrm{Tr}[(H_1-z)^{-1}-(H_2-z)^{-1}]= -\int_{\mathbb{R}} d\lambda\,\xi_{12}(\lambda)(\lambda-z)^{-2}, \quad z \in \rho(H_1) \cap \rho(H_2). \tag{5.3}$$

Moreover if $S_{12}(\lambda)$ denotes the on-shell scattering operator associated with (H_1,H_2) then

$$\det S_{12}(\lambda)=e^{-2\pi i \xi_{12}(\lambda)} \quad \text{for a.e. } \lambda \in \sigma_{ac}(H_2). \tag{5.4}$$

(b) Assume H(ii). Then

$$\mathrm{Tr}[(H_1-z)^{-1}-(H_2-z)^{-1}]= -\frac{d}{dz}\ln\det[1+u_{12}(H_2-z)^{-1}v_{12}], \quad z \in \rho(H_1) \cap \rho(H_2). \tag{5.5}$$

Lemma 5.1 (a) is a standard result (cf. e.g. [3], [29], [46]). Part (b) follows from [23], [34]. Next we introduce

H(iv). Assume that ξ_{12} is bounded and piecewise continuous on $\mathbb{R}$ and $\xi_{12}(\lambda)=0$ for $\lambda<0$.

Lemma 5.2. (a) Assume H(iii) and $(1+|.|)^{-1}\xi_{12} \in L^1(\mathbb{R})$. Then

$$\int_{\mathbb{R}} d\lambda\,\xi_{12}(\lambda)(\lambda-z)^{-1} = \ln\det[1+u_{12}(H_2-z)^{-1}v_{12}], \quad z \in \rho(H_1) \cap \rho(H_2). \tag{5.6}$$

(b) Assume H(iv) in addition to the conditions in (a). Then

$$[\xi_{12}(\lambda_+)+\xi_{12}(\lambda_-)]/2= \frac{1}{2\pi i}\lim_{\varepsilon\downarrow 0} \ln \frac{\det[1+u_{12}(H_2-\lambda-i\varepsilon)^{-1}v_{12}]}{\det[1+u_{12}(H_2-\lambda+i\varepsilon)^{-1}v_{12}]}, \quad \lambda \in \mathbb{R}. \tag{5.7}$$

Lemma 5.2 (a) immediately follows from Eqs. (5.3) and (5.5). Eq. (5.7) results from standard properties of the Poisson kernel ([9]).

Remark 5.3. The high-energy assumption H(iii) typically holds for one-dimensional systems but fails in general in dimensions ≥ 2. As shown in [5] one can introduce modified Fredholm determinants which overcome this difficulty.

Given the above preliminaries we now define Witten's (resolvent) regularized index $\Delta(z)$ ([13])

$$\Delta(z)=-z\mathrm{Tr}[(H_1-z)^{-1}-(H_2-z)^{-1}],\quad z\in\mathbb{C}\backslash[0,\infty) \tag{5.8}$$

and Witten's index Δ([50]) by

$$\Delta=\lim_{\substack{z\to 0\\ |\mathrm{Re}z|\leq c_0|\mathrm{Im}z|}}\Delta(z) \tag{5.9}$$

(for some $c_0>0$) whenever the limit exists. Instead of Eq. (5.8) one could as well consider a heat-kernel regularization but we omit the details. The anomaly A is then defined by ([8])

$$A=-\lim_{\substack{z\to\infty\\ |\mathrm{Re}z|\leq c_1|\mathrm{Im}z|}}\Delta(z) \tag{5.10}$$

(for some $c_1>0$) whenever the limit exists. We get ([5], [22])

Theorem 5.4. (a) Assume H(i) and suppose A to be Fredholm. Then

$$\Delta=i(A) \tag{5.11}$$

where

$$i(A)=\dim\mathrm{Ker}(A)-\dim\mathrm{Ker}(A^*)$$
$$(=\dim\mathrm{Ker}(H_1)-\dim\mathrm{Ker}(H_2)) \tag{5.12}$$

denotes the Fredholm index of A.

(b) Assume H(i) and H(iv). Then

$$\Delta=-\xi_{12}(0_+). \tag{5.13}$$

If in addition $\lim_{\lambda\to\infty}\xi_{12}(\lambda)=\xi_{12}(\infty)$ exists then

$$A=\xi_{12}(\infty). \tag{5.14}$$

Part (a) follows from the Laurent expansions of $(H_j-z)^{-1}$ near $z=0$ ([30]). We also note that A is Fredholm iff $\inf\sigma_{ess}(H_1)>0$. Part (b) results from Eqs. (5.3) and (5.8) ([5]).

Next we describe an important invariance property of $\Delta(z)$ under sufficiently small perturbations of A. Let B be another closed operator in $\mathcal{H}$ infinitesimally bounded w.r.t. to A and define on $\mathcal{D}(A)$

$$A_\beta=A+\beta B,\quad \beta\in\mathbb{R}. \tag{5.15}$$

In obvious notation the quantities $H_{1,\beta}$, $H_{2,\beta}$, $u_{12,\beta}$, $v_{12,\beta}$, $\xi_{12,\beta}$ and

$\Delta(\beta,z)$ then result after replacing A by A_β. Then we have ([22])

Theorem 5.5. Let $z_o \in \mathbb{C}\setminus[0,\infty)$ and assume

(i) $(H_{1,\beta}-z_o)^{-1}-(H_{2,\beta}-z_o)^{-1} \in \mathcal{B}_1(\mathcal{H})$ for all $\beta \in \mathbb{R}$.

(ii) $B^*B(H_1-z_o)^{-1}$, $BB^*(H_2-z_o)^{-1} \in \mathcal{B}_\infty(\mathcal{H})$,

$$[A^*B+B^*A](H_1-z_o)^{-1},\ [AB^*+BA^*](H_2-z_o)^{-1} \in \mathcal{B}_\infty(\mathcal{H}).$$

($\mathcal{B}_\infty(\mathcal{H})$ the set of compact operators in $\mathcal{H}$).

(iii) $(H_1-z_o)^{-1}B^*B(H_1-z_o)^{-1}$, $(H_2-z_o)^{-1}BB^*(H_2-z_o)^{-1} \in \mathcal{B}_1(\mathcal{H})$,

$(H_1-z_o)^{-1}[A^*B+B^*A](H_1-z_o)^{-1}$, $(H_2-z_o)^{-1}[AB^*+BA^*](H_2-z_o)^{-1} \in \mathcal{B}_1(\mathcal{H})$.

(iv) $(H_1-z_o)^{-M}B^*(H_2-z_o)^{-M} \in \mathcal{B}_1(\mathcal{H})$ for some $M \in \mathbb{N}$.

Then

$$\Delta(\beta,z)=\Delta(z),\ z\in\mathbb{C}\setminus[0,\infty),\ \beta\in\mathbb{R}. \tag{5.16}$$

For a more general result (where A acts between different Hilbert spaces $\mathcal{H}$ and $\mathcal{H}'$) see [22]. Eq. (5.16) can be derived as follows. Define

$$F(\beta,z)=\mathrm{Tr}[(H_{1,\beta}-z)^{-1}-(H_{2,\beta}-z)^{-1}],\ z\in\mathbb{C}\setminus[0,\infty),\ \beta\in\mathbb{R}. \tag{5.17}$$

Then conditions (i)-(iii) imply that

$$(\partial/\partial\beta)F(\beta,z)=-\mathrm{Tr}\{(H_{1,\beta}-z)^{-1}[A_\beta^*B+B^*A_\beta](H_{1,\beta}-z)^{-1}$$
$$-(H_{2,\beta}-z)^{-1}[A_\beta B^*+BA_\beta^*](H_{2,\beta}-z)^{-1}\}.$$

Using the commutation formulas ([17])

$$(A_\beta^*A_\beta-z)^{-1}A_\beta^* \subseteq A_\beta^*(A_\beta A_\beta^*-z)^{-1},$$
$$(A_\beta A_\beta^*-z)^{-1}A_\beta \subseteq A_\beta(A_\beta^*A_\beta-z)^{-1},\ z\in\mathbb{C}\setminus[0,\infty) \tag{5.18}$$

conditions (iii), (iv) and cyclicity of the trace then actually yield

$$(\partial/\partial\beta)F(\beta,z)=0,\ z\in\mathbb{C}\setminus[0,\infty),\ \beta\in\mathbb{R} \tag{5.19}$$

implying $F(\beta,z)=F(0,z)$ and hence Eq. (5.16).

Remark 5.6. Then result (5.16) yields the topological invariance of the regularized index $\Delta(z)$ in the examples at the end of this section. Moreover it proves the topological invariance of Δ and $\mathcal{A}$ whenever the limits in Eqs. (5.9) and (5.10) exist. In the case where A is Fredholm, the invariance of the Fredholm index i(A) (and thus of Δ by Eq. (5.11)) under relatively compact perturbations B w.r. to A, i.e.,

$$i(A+\beta B)=i(A),\ \beta\in\mathbb{R} \tag{5.20}$$

is a well known result ([30]). Eq. (5.16) does not assume A to be Fredholm but needs much stronger assumptions on B than just relative compactness.

Finally we note the corresponding invariance of Krein's spectral shift function ([22]).

Theorem 5.7. Assume H(iii) and H(iv) with A replaced by A_β and $(1+|\cdot|)^{-1}\xi_{12,\beta} \in L^1(\mathbb{R})$ for all $\beta \in \mathbb{R}$. Suppose in addition conditions (ii)-(iv) of Theorem 5.5. Then

$$\xi_{12,\beta}(\lambda) = \xi_{12}(\lambda) \quad \text{for all } \beta, \lambda \in \mathbb{R}. \tag{5.21}$$

Eq. (5.21) follows immediately from Eq. (5.7) and the invariance of the Fredholm determinants themselves

$$\det[1+u_{12,\beta}(H_{2,\beta}-z)^{-1}v_{12,\beta}] = \det[1+u_{12}(H_2-z)^{-1}v_{12}], \quad z \in \mathbb{C}\backslash[0,\infty) \tag{5.22}$$

which in turn is a consequence of Eqs. (5.5), (5.8) and (5.16).

Given this abstract approach we now turn to a series of one-dimensional examples taken from [5].

Example 5.8. Let $\mathcal{H} = L^2(\mathbb{R})$ and

$$A = \left(\frac{d}{dx} + \phi\right)\Big|_{H^{2,1}(\mathbb{R})} \tag{5.23}$$

where ϕ fulfills

$$\begin{aligned} &\phi, \phi' \in L^\infty(\mathbb{R}) \text{ are real-valued,} \\ &\lim_{x\to\pm\infty} \phi(x) = \phi_\pm \in \mathbb{R}, \quad \phi_-^2 \leq \phi_+^2, \\ &\int_{\mathbb{R}} dx(1+|x|^2)|\phi'(x)| < \infty, \quad \pm\int_0^{\pm\infty} dx(1+|x|^2)|\phi(x)-\phi_\pm| < \infty. \end{aligned} \tag{5.24}$$

In this case H_1 and H_2 explicitly read

$$H_j = \left(-\frac{d^2}{dx^2} + \phi^2 + (-1)^j\phi'\right)\Big|_{H^{2,2}(\mathbb{R})}, \quad j=1,2. \tag{5.25}$$

Then

$$\Delta(z) = [\phi_+(\phi_+^2-z)^{-1/2} - \phi_-(\phi_-^2-z)^{-1/2}]/2, \quad z \in \mathbb{C}\backslash[0,\infty) \tag{5.26}$$

and hence

$$\Delta = \begin{cases} [\mathrm{sgn}(\phi_+)-\mathrm{sgn}(\phi_-)]/2, & \phi_-\neq 0, \ \phi_+\neq 0 \\ [\mathrm{sgn}(\phi_+)]/2, & \phi_-=0, \ \phi_+\neq 0 \\ 0, & \phi_-=\phi_+ \end{cases}, \quad A=0 \tag{5.27}$$

$$\begin{aligned} \xi_{12}(\lambda) = {}& \pi^{-1}\{\theta(\lambda-\phi_+^2)\arctan[(\lambda-\phi_+^2)^{1/2}/\phi_+] - \theta(\lambda-\phi_-^2)\arctan[(\lambda-\phi_-^2)^{1/2}/\phi_-]\} \\ & +\theta(\lambda)[\mathrm{sgn}(\phi_-)-\mathrm{sgn}(\phi_+)]/2, \quad \phi_-\neq 0, \ \phi_+\neq 0, \\ \xi_{12}(\lambda) = {}& \pi^{-1}\theta(\lambda-\phi_+^2)\arctan[(\lambda-\phi_+^2)^{1/2}/\phi_+] - \theta(\lambda)[\mathrm{sgn}(\phi_+)]/2, \quad \phi_-=0, \ \phi_+\neq 0, \end{aligned} \tag{5.28}$$

$\xi_{12}(\lambda)=0,\ \phi_-=\phi_+;\ \lambda\in\mathbb{R}$.

(Here $\theta(x)=1$ for $x\geq 0$ and $\theta(x)=0$ for $x<0$). Eqs. (5.26)-(5.28) clearly demonstrate the topological invariance of these quantities as discussed before since they only depend on the asymptotic values $\phi_\pm$ of $\phi(x)$ and not on its local properties. In fact replace $\phi(x)$ by $\phi(x)+\beta\psi(x)$, $\beta\in\mathbb{R}$ where

$\psi,\psi'\in L^\infty(\mathbb{R})$ are real-valued

$$\psi(x),\psi'(x)=O(|x|^{-3-\varepsilon}) \text{ for some } \varepsilon>0 \text{ as } |x|\to\infty. \tag{5.29}$$

Then the perturbation B (cf. Eq. (5.15)) given by multiplication with ψ leaves the regularized index invariant since the hypotheses of Theorem 5.5 are satisfied ([22]).

Concerning zero-energy properties of H_j, $j=1,2$ we note

	zero-energy resonance of H_1	of H_2	$\sigma_p(H_1)\cap\{0\}$	$\sigma_p(H_2)\cap\{0\}$	Δ	$i(A)$
$\phi_-<0<\phi_+$	No	No	$\{0\}$	$\emptyset$	1	1
$\phi_+<0<\phi_-$	No	No	$\emptyset$	$\{0\}$	-1	-1
$\phi_+,\phi_->0$ or $\phi_+,\phi_-<0$	No	No	$\emptyset$	$\emptyset$	0	0
$\phi_-=0,\phi_+\neq 0$	Yes	No	$\emptyset$	$\emptyset$	1/2	
$\phi_-=\phi_+=0$	Yes	Yes	$\emptyset$	$\emptyset$	0	

These zero-energy results easily follow from the fact that

$$Af=0,\ A^*g=0 \tag{5.30}$$

has the solutions

$$f(x)=f(0)e^{-\int_0^x dt\,\phi(t)}=O(e^{-\phi_\pm x}) \text{ as } x\to\pm\infty,$$

$$g(x)=g(0)e^{\int_0^x dt\,\phi(t)}=O(e^{\phi_\pm x}) \text{ as } x\to\pm\infty. \tag{5.31}$$

In order do derive Eq. (5.26) we introduce

$$V_j(x)=\phi^2(x)+(-1)^j\phi'(x),\ j=1,2,\quad V_\pm=\phi_\pm^2,$$

$$V_{12}(x)=-2\phi'(x) \tag{5.32}$$

and apply the end of Sect. 2. In particular we combine Eqs. (2.59) and (5.5) to get

$$\mathrm{Tr}[(H_1-z)^{-1}-(H_2-z)^{-1}]=\frac{d}{dz}\ln\frac{W(f_{2-}(z),f_{2+}(z))}{W(f_{1-}(z),f_{1+}(z))}\,,\quad z\in\mathbb{C}\backslash[0,\infty). \tag{5.33}$$

In addition we note that ([17], [44])

$$H_1 f=zf,\ z\in\mathbb{C}\backslash\{0\}\ \text{implies}\ H_2(Af)=z(Af),$$

$$H_2 g=z'g,\ z'\in\mathbb{C}\backslash\{0\}\ \text{implies}\ H_1(A^*g)=z'(A^*g) \tag{5.34}$$

where $f(g)$ is a distributional solution of $H_1(H_2)$. Assuming now that $f_{1\pm}(z,x)$, $z\neq 0$ are Jost solutions of H_1 according to Eq. (2.2) then

$$\begin{cases} f_{1\pm}(z,x) \\ f_{2\pm}(z,x)=(\pm ik_\pm+\phi_\pm)^{-1}(Af_{1\pm})(z,x), \quad z\neq 0 \end{cases} \tag{5.35}$$

are correctly normalized Jost solutions for (H_1,H_2). Thus Eq. (2.59) yields

$$\det[1-2|\phi'|^{1/2}\mathrm{sgn}(\phi')(H_2-z)^{-1}|\phi'|^{1/2}]$$

$$=(-ik_-+\phi_-)(ik_++\phi_+)W(f_{1-}(z),f_{1+}(z))/W((Af_{1-})(z),\ (Af_{1+})(z)),$$

$$z\in\mathbb{C}\backslash[0,\infty). \tag{5.36}$$

Finally we note the simple identity

$$W((Af)(z),\ (A\tilde{f})(z))=zW(f(z),\tilde{f}(z)),\quad z\in\mathbb{C} \tag{5.37}$$

where f, $\tilde{f}$ are distributional solutions of

$$(H_1\eta(z))(x)=z\eta(z,x),\quad z\in\mathbb{C}. \tag{5.38}$$

Thus Eq. (5.36) becomes

$$\det[1-2|\phi'|^{1/2}\mathrm{sgn}(\phi')(H_2-z)^{-1}|\phi'|^{1/2}]$$

$$=W(f_{1-}(z),f_{1+}(z))/W(f_{2-}(z),f_{2+}(z))=(-ik_-+\phi_-)(ik_++\phi_+)/z,$$

$$z\in\mathbb{C}\backslash[0,\infty) \tag{5.39}$$

and hence Eq. (5.26) results from Eqs. (5.8) and (5.33).

The result (5.26) has first been derived in [13] and since then been rederived by numerous authors [1], [7], [25], [26], [27], [35], [39], [40] and [47].

If $\phi_-\in\mathbb{R}$ and $\lim_{x\to+\infty}\phi(x)=+\infty$ (i.e. $\phi_+=\infty$) one can follow the above steps using Sect. 3 to derive

$$\Delta(z)=[1-\phi_-(\phi_-^2-z)^{-1/2}]/2, \quad z\in\mathbb{C}\backslash[0,\infty),$$

$$\xi_{12}(\lambda)=\begin{cases} -\pi^{-1}\theta(\lambda-\phi_-^2)\arctan[(\lambda-\phi_-^2)^{1/2}/\phi_-] \\ +\theta(\lambda)[\mathrm{sgn}(\phi_-)-1]/2, \quad \phi_-\neq 0 \\ -\theta(\lambda)/2, \quad \phi_-=0. \end{cases} \tag{5.40}$$

This case contains supersymmetric Liouville quantum meachnics ([40]).

Next we discuss examples on the half-line $(0,\infty)$.

Example 5.9. Let $\mathcal{H}=L^2(0,\infty)$ and

$$A=\left(\frac{d}{dr}+\tilde{\phi}(r)\right)\Big|_{H_o^{2,1}(0,\infty)} \tag{5.41}$$

where $\tilde{\phi}$ fulfills

$$\tilde{\phi},\tilde{\phi}'\in L^\infty(0,\infty) \text{ are real-valued,}$$

$$\lim_{r\to\infty}\tilde{\phi}(r)=\tilde{\phi}_+\in\mathbb{R}, \quad \lim_{r\to 0_+}\tilde{\phi}(r)=\tilde{\phi}_o\in\mathbb{R}, \tag{5.42}$$

$$\int_o^\infty dr\, r(1+r)|\tilde{\phi}'(r)|<\infty, \quad \int_o^\infty dr\, r(1+r)|\tilde{\phi}(r)-\tilde{\phi}_+|<\infty.$$

In this case H_1 and H_2 read

$$H_1=\left(-\frac{d^2}{dr^2}+\tilde{\phi}^2-\tilde{\phi}'\right)_F \tag{5.43}$$

where "F" denotes the Friedrichs extension of the corresponding operator restricted to $C_o^\infty(0,\infty)$ and

$$H_2=-\frac{d^2}{dr^2}+\tilde{\phi}^2+\tilde{\phi}',$$

$$\mathcal{D}(H_2)=\{g\in L^2(0,\infty)\,|\,g,\ g'\in AC_{loc}(0,\infty);\ g'(0_+)-\tilde{\phi}_o g(0_+)=0;\ g''\in L^2(0,\infty)\}. \tag{5.44}$$

Then

$$\Delta(z)=(z/2)(\tilde{\phi}_+^2-z)^{-1/2}[\tilde{\phi}_++(\tilde{\phi}_+^2-z)^{1/2}]^{-1}, \quad z\in\mathbb{C}\backslash[0,\infty) \tag{5.45}$$

and hence

$$\Delta=\begin{cases} -[1-\mathrm{sgn}(\tilde{\phi}_+)]/2, \quad \tilde{\phi}_+\neq 0 \\ -1/2, \quad \tilde{\phi}_+=0 \end{cases}, \quad A=1/2, \tag{5.46}$$

$$\xi_{12}(\lambda)=\begin{cases} \pi^{-1}\theta(\lambda-\tilde{\phi}_+^2)\arctan[(\lambda-\tilde{\phi}_+^2)^{1/2}/\tilde{\phi}_+]+\theta(\lambda)\theta(-\tilde{\phi}_+), \quad \tilde{\phi}_+\neq 0, \\ \theta(\lambda)/2, \quad \tilde{\phi}_+=0; \quad \lambda\in\mathbb{R}. \end{cases} \tag{5.47}$$

Again Eqs. (5.45)-(5.47) exhibit the topological invariance of all these quantities since only $\tilde{\phi}_+$ enters. Concerning zero-energy properties we note

	zero-energy of H_1	resonance of H_2	$\sigma_p(H_1)\cap\{0\}$	$\sigma_p(H_2)\cap\{0\}$	Δ	$i(A)$
$\tilde{\phi}_+>0$	No	No	$\emptyset$	$\emptyset$	0	0
$\tilde{\phi}_+<0$	No	No	$\emptyset$	$\{0\}$	-1	-1
$\tilde{\phi}_+=0$	No	Yes	$\emptyset$	$\emptyset$	-1/2	

In order to derive Eq. (5.45) we introduce Jost solutions

$$f_{j\pm}(z,r)=e^{\pm ik_+r}-\int_r^\infty dr'k_+^{-1}\sin[k_+(r-r')][\tilde{\phi}^2(r')-\tilde{\phi}_+^2+(-1)^j\tilde{\phi}'(r')]f_{j\pm}(z,r'),$$
$$z\in\mathbb{C},\ j=1,2 \qquad (5.48)$$

where

$$k_+(z)=(z-\tilde{\phi}_+^2)^{1/2},\ \operatorname{Im}k_+(z)\geq 0,\quad z\in\mathbb{C} \qquad (5.49)$$

and regular solutions

$$\psi_1(z,r)=k_+^{-1}\sin k_+r+\int_0^r dr'k_+^{-1}\sin[k_+(r-r')][\tilde{\phi}^2(r')-\tilde{\phi}_+^2-\tilde{\phi}'(r')]\psi_1(z,r'),$$
$$\psi_2(z,r)=\cos k_+r+\tilde{\phi}_0k_+^{-1}\sin k_+r \qquad (5.50)$$
$$+\int_0^r dr'k_+^{-1}\sin[k_+(r-r')][\tilde{\phi}^2(r')-\tilde{\phi}_+^2+\tilde{\phi}'(r')]\psi_2(z,r'),\ z\in\mathbb{C}.$$

Using again property (5.34) we assume that $f_{1\pm}(z,r)$, $z\neq 0$ is normalized according to Eq. (5.48). Then

$$\begin{cases} f_{1\pm}(z,r), \\ f_{2\pm}(z,r)=(\pm ik_\pm+\tilde{\phi}_+)^{-1}(Af_{1\pm})(z,r),\ z\neq 0 \end{cases} \qquad (5.51)$$

are correctly normalized Jost functions for (H_1,H_2). Similarly assume that $\psi_1(z,r)$, $z\neq 0$ fulfills Eq. (5.50) then

$$\begin{cases} \psi_1(z,r) \\ \psi_2(z,r)=(A\psi_1)(z,r),\ z\neq 0 \end{cases} \qquad (5.52)$$

are correctly normalized regular solutions for (H_1,H_2). The rest is now identical to Example 5.8. First of all one derives as in Eq. (5.33) (cf. e.g. [11], [37])

$$\mathrm{Tr}[(H_1-z)^{-1}-(H_2-z)^{-1}] = \frac{d}{dz}\ln\frac{W(\psi_2(z),f_{2+}(z))}{W(\psi_1(z),f_{1+}(z))}\ ,\quad z\in\mathbb{C}\backslash[0,\infty) \tag{5.53}$$

and then one calculates as in Eq. (5.37) that

$$W(A\psi_1)(z),\ (Af_{1+})(z))=zW(\psi_1(z),f_{1+}(z)),\ z\in\mathbb{C}. \tag{5.54}$$

Next we consider a generalization which allows to discuss n-dimensional spherically symmetric systems (cf. e.g. [28], [32]).

<u>Example 5.10.</u> Let $\mathcal{H}=L^2(0,\infty)$ and

$$A=\overline{(\frac{d}{dr}+\phi)\Big|_{C_0^\infty(0,\infty)}} \tag{5.55}$$

where ϕ fulfills

$$\phi(r)=\phi_0 r^{-1}+\tilde{\phi}(r),\ \phi_0\leq -1/2,\ r>0,$$

$$\tilde{\phi},\tilde{\phi}'\in L^\infty(0,\infty)\text{ are real-valued,}$$

$$\lim_{r\to\infty}\tilde{\phi}(r)=\tilde{\phi}_+\in\mathbb{R}, \tag{5.56}$$

$$\int_0^\infty dr\,\omega_{\phi_0}(r)[|\tilde{\phi}'(r)|+r^{-1}|\tilde{\phi}(r)-\tilde{\phi}_+|]<\infty,$$

$$\int_0^\infty dr\,\omega_{\phi_0}(r)|\tilde{\phi}(r)-\tilde{\phi}_+|<\infty$$

and the weight function ω_{ϕ_0} is defined by

$$\omega_{\phi_0}(r)=\begin{cases} r(1+r) & \text{if } \phi_0<-1/2 \\ r(1+|\ln r|),\ 0<r\leq 1/2 & \\ r(1+r),\ r\geq 1/2 & \text{if } \phi_0=-1/2. \end{cases} \tag{5.57}$$

Now H_1 and H_2 are given by

$$H_j=(-\frac{d^2}{dr^2}+\phi^2+(-1)^j\phi')_F\ ,\ j=1,2. \tag{5.58}$$

Explicitly we have

$$\phi^2(r)\mp\phi'(r)=(\phi_0^2\pm\phi_0)r^{-2}+2\phi_0\tilde{\phi}_+r^{-1}+\tilde{\phi}_+^2$$
$$+\tilde{\phi}^2(r)-\tilde{\phi}_+^2\mp\tilde{\phi}'(r)+2\phi_0[\tilde{\phi}(r)-\tilde{\phi}_+]r^{-1},\ r>0. \tag{5.59}$$

Then

$$\Delta(z)=(z/2)(\tilde{\phi}_+^2-z)^{-1/2}[\tilde{\phi}_+-(\tilde{\phi}_+^2-z)^{1/2}]^{-1},\ z\in\mathbb{C}\backslash[0,\infty) \tag{5.60}$$

and hence

$$\Delta = \begin{cases} [1+\mathrm{sgn}(\tilde{\phi}_+)]/2, & \tilde{\phi}_+ \neq 0 \\ & , A=-1/2, \\ 1/2, & \tilde{\phi}_+ = 0 \end{cases} \tag{5.61}$$

$$\xi_{12}(\lambda) = \begin{cases} \pi^{-1}\theta(\lambda-\tilde{\phi}_+^2)\arctan[(\lambda-\tilde{\phi}_+^2)^{1/2}/\tilde{\phi}_+]-\theta(\lambda)\theta(\tilde{\phi}_+), & \tilde{\phi}_+ \neq 0, \\ -\theta(\lambda)/2, & \tilde{\phi}_+=0; \ \lambda \in \mathbb{R}. \end{cases} \tag{5.62}$$

The topological invariance in Eqs. (5.60)-(5.62) is obvious. We also note

	zero-energy resonance of H_1	of H_2	$\sigma_p(H_1) \cap \{0\}$	$\sigma_p(H_2) \cap \{0\}$	Δ	$i(A)$
$\tilde{\phi}_+ > 0$	No	No	$\{0\}$	$\emptyset$	1	1
$\tilde{\phi}_+ < 0$	No	No	$\emptyset$	$\emptyset$	0	0
$\tilde{\phi}_+ = 0$	No	No	$\emptyset$	$\emptyset$	1/2	

If $\tilde{\phi}_+=0$ the result $\Delta=1/2$ is not due to a zero-energy (threshold) resonance but due to the long-range nature of the relative interaction $V_{12}(r)= =2\phi_o r^{-2}+o(r^{-2})$ as $r\to\infty$. Since Eq. (5.60) is independent of ϕ_o this result holds in any dimension ≥ 2 and for any value of the angular momentum.

In order to derive Eq. (5.60) one could follow the strategy of Example 5.9 step by step since formula (5.53) for suitably normalized Jost- and regular solutions remains valid in the present case (although we are dealing with a long-range problem!). To shorten the presentation we will instead use a different approach based on the topological invariance property of $\Delta(z)$ and $\xi_{12}(\lambda)$ (this approach obviously also works in Example 5.9): Because of Theorem 5.7 it suffices to choose $\tilde{\phi}(r)=\tilde{\phi}_+, r\geq 0$ in Example 5.10. Then

$$H_j = \left(-\frac{d^2}{dr^2}+[\phi_o^2-(-1)^j\phi_o]r^{-2}+2\phi_o\tilde{\phi}_+ r^{-1}+\tilde{\phi}_+^2\right)_F, \quad j=1,2 \tag{5.63}$$

and ([21])

$$S_j(\lambda) = \frac{\Gamma(2^{-1}+2^{-1}(-1)^j-\phi_o+i(\phi_o\tilde{\phi}_+/k_+))}{\Gamma(2^{-1}+2^{-1}(-1)^j-\phi_o-i(\phi_o\tilde{\phi}_+/k_+))} e^{i\pi[2^{-1}-(-1)^j 2^{-1}+\phi_o]}, \tag{5.64}$$

$$\lambda > \tilde{\phi}_+^2, \quad j=1,2$$

($k_+(\lambda)$ defined in Eq. (5.49)) implying

$$S_{12}(\lambda)=S_1(\lambda)S_2(\lambda)^{-1}=(\tilde{\phi}_+-ik_+)/(\tilde{\phi}_++ik_+),\quad \lambda>\tilde{\phi}_+^2. \tag{5.65}$$

Eq. (5.65) proves Eq. (5.62). Now Eq. (5.60) follows by explicit integration ([24], p. 556) in Eq. (5.3).

The result (5.60) in the special case $\tilde{\phi}(r)\equiv 0$ has been discussed in [1], [2] by different means.

Finally we briefly discuss nonlocal interactions.

Example 5.11. Let $\mathcal{H}=L^2(0,\infty)$ and

$$A=\frac{d}{dr}\Big|_{H_0^{2,1}(0,\infty)}+B \tag{5.66}$$

where

$$B,A^*B,AB^*\in \mathcal{B}_1(L^2(0,\infty)). \tag{5.67}$$

In this case the assumptions of Theorem 5.5 are trivially fulfilled and hence Eqs. (5.45)-(5.47) in the special case $\phi(r)\equiv 0$ hold. In particular

$$\Delta(z)=\Delta=-1/2,\quad z\in\mathbb{C}\backslash[0,\infty). \tag{5.68}$$

In order to illustrate the possible complexity of zero-energy properties of H_1 and H_2 despite the simplicity of Eq. (5.68) it suffices to treat the following rank two example:

$$B=\alpha(f,.)f+\beta(g,.)g,\quad \alpha,\beta\in\mathbb{R}$$
$$f,g\in C_0^1(0,\infty),\quad f\geq 0,\quad g\geq 0,\quad f\neq g. \tag{5.69}$$

Straightforward calculations then yield

	zero-energy resonance of H_1	zero-energy resonance of H_2	$\sigma_p(H_1)\cap\{0\}$	$\sigma_p(H_2)\cap\{0\}$	Δ
Case I	No	Yes	$\emptyset$	$\emptyset$	-1/2
Case II	Yes	No	$\emptyset$	$\{0\}$	-1/2
Case III	No	Yes	$\{0\}$	$\{0\}$	-1/2

Here the following case distinction has been used:

Case I. $\Psi(\alpha,\beta)\neq 0$.

Case II. $\Psi(\alpha,\beta)=0,\ \alpha\neq 2G(\infty)\{F(\infty)[(f,G)-(g,F)]\}^{-1}$.

Case III. $\Psi(\alpha,\beta)=0,\ \alpha=2G(\infty)\{F(\infty)[(f,G)-(g,F)]\}^{-1}$.

where

$$F(x)=\int_0^x dx' f(x'),\quad G(x)=\int_0^x dx' g(x'),$$

$$\Psi(\alpha,\beta)= [1+\alpha(f,F)][1+\beta(g,G)]-\alpha\beta(f,G)(g,F). \tag{5.70}$$

Acknowledgements

I am indebted to D.Bollé, H.Grosse, W.Schweiger and B.Simon for numerous discussions and joint collaborations which led to most of the results presented above.

It is a great pleasure to thank E.Balslev for all his efforts to organize this conference and all participants for creating such a stimulating atmosphere.

This work is part of Research Project No. P5588 supported by Fonds zur wissenschaftlichen Forschung in Österreich.

References

1. R.Akhoury and A.Comtet, Nucl. Phys. B246 (1984), 253-278.
2. N.A.Alves, H.Aratyn and A.H.Zimmerman, Phys. Rev. D31 (1985), 3298-3300.
3. H.Baumgärtel and M.Wollenberg, "Mathematical Scattering Theory", Birkhäuser, Basel, 1983.
4. M.S.Birman and M.G.Krein, Sov. Math. Dokl. 3 (1962), 740-744.
5. D.Bollé, F.Gesztesy, H.Grosse and W.Schweiger, paper in preparation.
6. P.J.M.Bongaarts and S.N.M.Ruijsenaars, Ann. Inst. H.Poincaré A26 (1977), 1-17.
7. D.Boyanovsky and R.Blankenbecler, Phys. Rev. D30 (1984), 1821-1824.
8. D.Boyanovsky and R.Blankenbecler, Phys. Rev. D31 (1985), 3234-3250.
9. H.Bremermann, "Distributions, Complex Variables and Fourier Transforms", Addison-Wesley, New York, 1965.
10. V.S.Buslaev, "Theoret. Math. Phys. 58 (1984), 153-159.
11. V.S.Buslaev and L.D.Faddeev, Sov. Math.Dokl. 1 (1960), 451-454.
12. E.Caliceti, Ann. Inst. H.Poincaré A42 (1985), 235-251.
13. C.Callias, Commun. Math. Phys. 62 (1978), 213-234.
14. K.Chadan and P.C.Sabatier, "Inverse Problems in Quantum Scattering Theory", Springer, New York, 1977.
15. A.Cohen and T.Kappeler, Indiana Univ. Math. J. 34 (1985), 127-180.
16. E.B.Davies and B.Simon, Commun. Math. Phys. 63 (1978), 277-301.
17. P.A.Deift, Duke Math. J. 45 (1978), 267-310.
18. P.Deift and E.Trubowitz, Commun. Pure Appl. Math. 32 (1979), 121-251.
19. N.E.Firsova, J. Sov. Math. 11 (1979), 487-497.
20. N.E.Firsova, Theoret. Math. Phys. 62 (1985), 130-140.
21. F.Gesztesy, W.Plessas and B.Thaller, J. Phys. A13 (1980), 2659-2671.

22. F.Gesztesy and B.Simon, paper in preparation.
23. I.C.Gohberg and M.G.Krein, "Introduction to the Theory of Linear Nonselfadjoint Operators", AMS, Providence, 1969.
24. I.S.Gradshteyn and I.M.Ryzhik, "Table of Integrals, Series, and Products", Academic, New York, 1965.
25. M.Hirayama, Progr. Theoret. Phys. 70 (1983), 1444-1453.
26. M.Hirayama, Phys. Lett. 156B (1985), 225-230.
27. C.Imbimbo and S.Mukhi, Nucl. Phys. B242 (1984), 81-92.
28. T.D.Imbo and U.P.Sukhatme, Phys. Rev. Lett. 54 (1985), 2184-2187.
29. A.Jensen and T.Kato, Commun. Part. Diff. Eqs. 3 (1978), 1165-1195.
30. T.Kato, "Perturbation Theory for Linear Operators", Springer, Berlin, 1980.
31. W.Kohn, Phys. Rev. 115 (1959), 809-821.
32. V.A.Kostelecky and M.M.Nieto, Phys. Rev. A32 (1985), 1293-1298.
33. M.G.Krein, Sov. Math. Dokl. 3 (1962), 707-710.
34. M.G.Krein, "Topics in Differential and Integral Equations and Operator Theory", I.Gohberg (ed.), Birkhäuser, Basel, 1983, p. 107.
35. J.Lott, Commun. Math. Phys. 93 (1984), 533-558.
36. R.G.Newton, J.Math. Phys. 21 (1980), 493-505.
37. R.G.Newton, "Scattering Theory of Waves and Particles", Springer, New York, 1982.
38. R.G.Newton, J. Math. Phys. 24 (1983), 2152-2162; 26 (1985), 311-316.
39. A.J.Niemi and G.W.Semenoff, "Fermion Number Fractionization in Quantum Field Theory" to appear in Phys. Rep. (1986).
40. A.J.Niemi and L.C.R.Wijewardhana, Phys. Lett. 138B (1984), 389-392.
41. D.B.Pearson, Helv. Phys. Acta 47 (1974), 249-264.
42. M.Reed and B.Simon, "Methods of Modern Mathematical Physics III. Scattering Theory", Academic, New York, 1979.
43. M.Reed and B.Simon, "Methods of Modern Mathematical Physics IV". Analysis of Operators", Academic, New York, 1978.
44. U.W.Schmincke, Proc. Roy. Soc. Edinburgh 80A (1978), 67-84.
45. B.Simon, "Trace Ideals and their Applications", Cambridge Univ. Press Cambridge, 1979.
46. K.B.Sinha, "On the theorem of M.G.Krein", Univ. Geneva preprint, 1975.
47. M.Stone, Ann. Phys. 155 (1984), 56-84.
48. E.C.Titchmarsh, "Eigenfunction Expansions associated with Second-Order Differential Equations II", Clarendon Press, Oxford, 1958.
49. J.Weidmann, Math. Z. 98 (1967), 268-302; 180 (1982), 423-427.
50. E.Witten, Nucl. Phys. B202 (1982), 253-316.
51. V.A.Zheludev, Topics in Math. Phys. 4 (1971), 55-75.

CLASSICAL LIMIT AND CANONICAL PERTURBATION THEORY

Sandro Graffi
Dipartimento di Matematica, Università di Bologna
I-40127 BOLOGNA, Italy

A mio padre nel suo 81°compleanno.

1. Introduction.

The purpose of this lecture is to describe a procedure, developed in collaboration with T.Paul [1], to simultaneously generate, for holomorphic perturbations of the d-dimensional, non resonant harmonic oscillator, the formal power series solution of both the quantum spectral problem, i.e. the Rayleigh-Schrödinger expansion, as well as of the classical integrability problem, i.e. the Birkhoff expansion. It consists in remarking that in the Bargmann representation the Schrödinger equation can be written as the Hamilton-Jacobi equation plus explicit corrections in powers of $\hbar$, so that the spectral problem can be looked at as the problem of finding a Birkhoff normal form. Thus it can be recursively solved by the Birkhoff transformation, i.e. the algorithm of classical, canonical perturbation theory, up to the minor variant of working with the Laurent expansion instead of the Fourier one. The role of the actions is played by $n_i\hbar$, $n_i = 0,1,\dots$, $i = 1,\dots d$ the quantum numbers, so that the classical case is included in a natural way at the limit $n_i \to \infty$, $\hbar \to 0$, $n_i\hbar = A_i > 0$, A_i being the canonical action variables for the classical harmonic oscillator, $i = 1,\dots,d$. In this way all quantum corrections to each term of the Birkhoff expansion are determined. As a by-product of this and of the Kolomogorov-Arnold-Moser theorem if V is a polynomial a quantization formula is obtained, valid to all orders in perturbation theory for all quantum eigenvalues such that the product $n\hbar$, $n = (n_1,\dots,n_d)$, coincides with a KAM torus.

To formalize the above sketch, let me first state some abbreviations used throughout this paper. If $z = (z_1,\dots,z_d) \in \mathbb{C}^d$, then $z^2 = \sum_{i=1}^{d} z_i^2$, $(n_1\hbar,\dots,n_d\hbar) = n\hbar$, $(\omega_1 z_1,\dots,\omega_d z_d) = z$, $\sum_{i=1}^{d} \omega_i^2 z_i^2 = \omega^2 z^2$, $|z| = |z_1| + \dots + |z_d|$, $z! = z_1! \dots z_d!$. If $z \to f(z) \in \mathbb{C}$ is analytic at z, its gradient is denoted by $(\nabla_z f)(z)$, and its partial derivatives $\left(\frac{\partial^{\mu_1+\dots+\mu_d}}{\partial z_1^{\mu_1} \dots \partial z_d^{\mu_d}} f\right)(z_1,\dots,z_d)$ are denoted by $(D_z^{\mu})(z)$.
If $(z_1,z_2) \in \mathbb{C}^d \times \mathbb{C}^d$, their scalar product is denoted by $\langle z_1,z_2 \rangle$.
Furthermore, let:

$$H_o(\omega) = \frac{1}{2} \sum_{i=1}^{d} (p_i^2 + \omega_i^2 q_i^2) \tag{1.1}$$

be the Hamiltonian of the d-dimensional harmonic oscillator, with non-resonant frequencies $\omega = (\omega_1,\dots,\omega_d)$ fulfilling a Diophantine condition, i.e., if $\nu \in \mathbb{Z}^d$

$$\langle \omega,\nu \rangle = 0 \text{ iff } \nu = 0, \quad |\langle \omega,\nu \rangle|^{-1} < C(\omega)|\nu|^{d+1} \tag{1.2}$$

for some $C(\omega) > 0$;

$$f_o(A,\omega) = \langle \omega,A \rangle \tag{1.3}$$

the canonical image $H_o(C^{-1}(A,\phi))$ of H_o under the standard action-ngle canonical trasformation mapping $(\mathbb{R}^2 \setminus \{0\})^d$ onto $\mathbb{R}_+^d \times \mathbb{T}^d$:

$$C^{-1}(A,\phi) := \begin{cases} p_i = - \sqrt{2\omega_i A_i}\ \sin\phi_i \\ q_i = \sqrt{2A_i/\omega_i}\cos\phi_i \end{cases} \tag{1.4}$$

$$F_o(z,R) = \langle \omega z, R\rangle - \omega z^2 \tag{1.5}$$

the image $H_o(C_o^{-1}(z,R))$ under the canonical bijection of $\mathbb{C}^d \cong \mathbb{R}^{2d}$ onto itself:

$$C_o^{-1}(z,R) := \begin{cases} a_j = \sqrt{2\omega_j}\ R_j \\ p_j = - i\ \sqrt{2\omega_j}\ z_j - \omega_j R_j \end{cases} \tag{1.6}$$

$V(q) = V(q_1,\ldots,q_d)$ polynomial of degree 2m, bounded below, which, without loss of generality, can be taken without terms of order ≤ 2;

$$H(\omega,\varepsilon) = H_o(\omega) + \varepsilon V \tag{1.7}$$

the perturbed Hamiltonian; $T_o(\hbar,\varepsilon)$ the Schrödinger operator corresponding to $H_o(\omega)$, i.e. the maximal operator in $L^2(\mathbb{R}^d)$ generated by the differential expression $(-\hbar^2\Delta + \omega^2 q^2)/2$. $T_o(\hbar,\omega)$ is self-adjoint, with discrete spectrum given by the simple eigenvalues

$$\lambda_n(\hbar) = \langle n,\omega\rangle\hbar + \hbar(\omega)/2,\ n\in(\mathbb{N}\cup\{0\})^d \tag{1.8}$$

$$T(\hbar,\varepsilon) = T_o + \varepsilon V,\ D(T(\hbar,\varepsilon)) = D(T_o)\quad D(V),\quad \varepsilon \geq 0 \tag{1.9}$$

the Schrödinger operator corresponding to $H(\omega,\varepsilon)$. $T(\hbar,\varepsilon)$ is self-adjoint with discrete spectrum: its eigenvalues are denoted by $\lambda_n(\hbar,\varepsilon)$; U the Bargmann transform[2] $\psi \to (U\psi)(z) \equiv f(z)$, $z\in\mathbb{C}^d$, $a \to \psi(a)\in L^2(\mathbb{R}^d)$, defined as

$$(U\psi)(z) = \pi^{-d/2}(\omega_1\ldots\omega_d)^{1/2}\int_{\mathbb{R}^d} e^{-(z^2+q^2+\sqrt{2}\langle z,q\rangle)/2\hbar}\,\psi(q)\,dq \tag{1.10}$$

which is a unitary map between $L^2(\mathbb{R}^d)$ and the Hilbert space F_d of all entire functions $z \to f(z)$ from $\mathbb{C}^d$ to $\mathbb{C}$ such that

$$\int_{\mathbb{R}^d} |f(z)|^2 e^{-|z|^2}\prod_{i=1}^d dx_i dy_i < +\infty,\ z_i = x_i + iy_i \tag{1.11}$$

$$P_o(\hbar,\omega) = U H_o(\omega)U^{-1},\ P(\hbar,\omega,\varepsilon) = U\,T(\hbar,\varepsilon)U^{-1},\ Z_i(\hbar,\omega) = U q_i U^{-1} \tag{1.12}$$

the unitary images in F_d of the operators T_o, T, q_i, respectively, given by the maximal operators in F_d generated by the differential expressions

$$\tilde{P}_o(\hbar,\omega) = \hbar < \omega z, \nabla_z > + \hbar|\omega|/2 \tag{1.13}$$

$$\tilde{P}(\hbar,\omega,\varepsilon) = \tilde{P}_o(\hbar,\omega) + \varepsilon V(\mathbb{Z}(\hbar,\omega)) \tag{1.14}$$

$$\mathbb{Z}(\hbar,\omega) = (z + \hbar \nabla_z)/\sqrt{2\omega} \tag{1.15}$$

By (1.13), (1.14) it is seen that, up to the additive constant $\hbar|\omega|/2$, the Schrödinger eigenvalue problem for $T(\hbar,\varepsilon)$ takes the form:

$$\hbar < \omega z, \nabla_z > \psi(z,E) + \varepsilon V(\mathbb{Z}(\hbar,\omega))\, \psi(z,E) = E\, \psi(z,E) \tag{1.16}$$

where $\psi(z,E) \in F_d$. The first result can now be stated. Although it has been proved in Ref.(1) for a certain class of entire functions V, for the sake of simplicity only the polynomial case will be considered here.

<u>Proposition 1.</u>

Let $E \in \mathbb{C}$, $\varepsilon \geq 0$, and let the family of functions $z \to W(z,E,\varepsilon)$ be locally holomophic in $\mathbb{C}^d$ for amy fixed (E,ε). Set, for $\ell = 1,2,\ldots$,

$$R_\ell(W(z,E,\varepsilon)) = \sum_{|t|=\ell+1}^{2\ell} D_q^{|t|} V(\nabla_z W(\cdot)/\sqrt{2\omega}) \cdot \sum{}^{*} \prod_{|\mu|=2}^{\ell+1} \left(\frac{D_z^\mu W(\cdot)}{\mu!\sqrt{2\omega}} \right)^{a_\mu} \frac{1}{a_\mu!} \tag{1.17}$$

where $\sum^*$ means summation over all non-negative integers a_μ such that $\sum_{|\mu|=2}^{\ell+1} a_\mu = |t| - \ell$, $\sum_{|\mu|=2}^{\ell+1} \mu_i a_\mu = t_i$, $i = 1,\ldots,d$. Then:

(1) $z \to W(z,E,\varepsilon)$ is a locally holomorphic solution of

$$< \omega z, \nabla_z > W(z,E,\varepsilon) - \omega z^2 + \varepsilon\, |V(\,{}_z W(z,E,\varepsilon)/\sqrt{2\omega}) + \sum_{\ell=1}^{2m} \hbar^\ell R_\ell(W(z,E,\varepsilon))| = E \tag{1.18}$$

if and only if $\psi(z,E,\varepsilon) = \exp\{(W(z,E,\varepsilon) - z^2)/2\hbar\}$ is a locally holomorphic solution of the Schrödinger equation (1.16).

(2) Let $\varepsilon = 0$, and $E = f_o(A,\omega) = \langle\omega, A\rangle$, $A \in \mathbb{C}^d$. Then equation (1.18) admits the solution

$$W_o(A,z) = \sum_{i=1}^{d} (A_i \operatorname{Log} z_i + z_i^2/2) = \sum_{i=1}^{d} W_i^o(A_i, z_i) \tag{1.19}$$

(Log z denotes the principal branch of $z \to \log z$) which is the generating function of the canonical trasformation $C_1(A,\phi) = (z,R)$

$$(z,R) := \begin{cases} z_i = \sqrt{A_i}\, e^{i\phi_i} \\ R_i = 2\sqrt{A_i}\, \cos \phi_i \end{cases} \tag{1.20}$$

mapping $F_o(z,R)$ into $f_o(A,\omega)$, i.e. $F_o(C_1^{-1}(A,\phi)) = f_o(A,\omega)$. If $A = n\hbar$, $n \in (\mathbb{N} \cup \{0\})^d$, the functions $\psi_n(n\hbar,z) = \exp\{(W_o(n\hbar,z) - z^2)/2\hbar\}$ are, up to normalization, all eigenfunctions of the quantum harmonic oscillators, corresponding to the eigenvalues $f_o(n\hbar,\omega) = \lambda_n(\hbar) - \hbar|\omega|/2$.

(3) For $\varepsilon > 0$ the Birkhoff transformation (see e.g. Gallavotti[3], § 5.10) recursively yields a formal power series solution of (1.18), of initial point $W_o(A,z)$, $f_o(A,\omega)$, namely: for any bounded sphere $\Omega \subset \mathbb{C}^d$, $k = 1,2,\ldots$, $\ell = 1,\ldots,2km$ there are uniquely determined functions $(A,z) \to \Phi_k(A,z)$, $(A,z) \to W_k^\ell(A,z)$, holomorphic in $\Omega \times \mathbb{C}^d \setminus \{0\}$ and $A \to P_k(A)$, $A \to Q_k^\ell(A)$, holomorphic in Ω such that, if

$$W_k(A,z,\hbar) = \Phi_k(A,z) + \sum_{\ell=1}^{2km} \hbar^\ell W_k^\ell(A,z) \quad , \ k \geq 1 \tag{1.21}$$

$$\lambda_k(A,\hbar) = P_k(A) + \sum_{\ell=1}^{2km} \hbar^\ell Q_k^\ell(A) \quad , \ k \geq 1 \tag{1.22}$$

then

$$W_N(A,z,\hbar,\varepsilon) = \sum_{n=o}^{N} W_k(A,z,\hbar)\varepsilon^k, \quad \lambda_N(A,\hbar,\varepsilon) = \sum_{k=o}^{N} \lambda_k(A,\hbar)\varepsilon^k \tag{1.23}$$

solve equation (1.18) up to order ε^{N+1}, i.e., for $N = 1,2,\ldots$

$$\langle \omega z, \nabla_z \rangle W_N(A,z,\hbar,\varepsilon) - \omega^2 z^2 + \tag{1.24}$$
$$+ \varepsilon\, [V(\nabla_z W_N(A,z,\hbar,\varepsilon) + \sum_{\ell=1}^{2m} \hbar^\ell R_\ell (W_N(A,z,\hbar,\varepsilon))] - \lambda_N(A,\hbar,\varepsilon) = O(\varepsilon^{N+1})$$

(4) If $A = n\hbar$, the function $\lambda_k(n\hbar,\hbar)$ coincides with the k-th coefficient of the Rayleigh-Schrödinger expansion near the eigenvalue $\lambda_n(\hbar)$ of

$T_o(\omega)$. If $\hbar \to 0$, $n \to \infty$, $n\hbar = A$, $W_N(A,z,\varepsilon) = \sum_{k=1}^{N} \Phi_k(A,z)\varepsilon^k$ solves the Hamilton - Jacobi equation:

$$F_o(z,\nabla_z W(A,z,\varepsilon)) - \omega z^2 + \varepsilon V(\nabla_z W(A,z,\varepsilon)) = \tag{1.25}$$

$= $ function independent of z up to order ε^{N+1}, $N = 1,2,\ldots$

i.e. it coincides with the generating function of the canonical transformation C_N which sends the classical Hamiltonian $F_o(z,R) + \varepsilon V(R/\sqrt{2\omega})$, canonically equivalent to $H(\omega,\varepsilon)$ via (1.6), into the Birkhoff normal form $\sum_{k=o}^{N} P_k(A)\varepsilon^k$ for all $N = 1,2,\ldots$, $P_o(A) = f_o(A,\omega)$. Therefore $P_k(A)$, $k = 1,2,\ldots$, is the k-th coefficient of the Birkhoff expansion.

To state the second result, a further assumption is necessary. Write the Hamiltonian $H(\omega,\varepsilon)$ in the action-angle variables of $H_o(\omega)$, given by (1.4):

$$H(A,\phi;\varepsilon) = H_o(C^{-1}(A,\phi)) + \varepsilon V(C^{-1}(A,\phi)) \equiv f_o(A,\omega) + \varepsilon V(A,\phi) \tag{1.26}$$

Let $A \to V_o(A)$ be the mean value of $V(A,\phi)$ over the torus $\mathbb{T}^d$, and:

$$M(A) = \| \partial^2 V_o(A)/\partial A_i \partial A_j \|_{i,j=1,\ldots,d} \tag{1.27}$$

$$S_\rho(A_o) = \{A \in \mathbb{C}^d : \max |A_i - A_i^o| \le \rho\}, \quad G(\rho,\Omega) = \bigcup_{A_o \in \Omega} S_\rho(A_o) \tag{1.28}$$

Then there is $\eta > 0$ such that

$$\sup_{A \in G(\rho,\Omega)} |\det M(A)|^{-1} < \eta < +\infty \tag{1.29}$$

Under this assumption the KAM theorem (see e.g. Chierchia-Gallavotti[4], Gallavotti[5]) yields, for ε suitably small, the existence of a subset $\Gamma^\infty(\varepsilon)$ of Ω (the "KAM tori"), such that its Lebesgue measure $\mu(\Gamma^\infty(\varepsilon)) = (1-K\varepsilon)\mu(\Omega)$ for some $K > 0$, and of $(A,\varepsilon) \to f^\infty(A,\varepsilon)$, defined on $\Gamma^\infty(\varepsilon)$, which admits a C^∞ extension to Ω, such that $H(\omega,\varepsilon)$ is canonically equivalent to $f^\infty(A,\varepsilon)$ on $\Gamma^\infty(\varepsilon)$. Hence $H(\omega,\varepsilon)$ is integrable on $\Gamma^\infty(\varepsilon)$.

Corollary 1. For amy $0 < \delta < +\infty$, and $\bar{\varepsilon} > 0$ suitably small, there is

$(A,\hbar,\varepsilon) \to g^{\infty}(A,\hbar,\varepsilon) \in C^{\infty}(\,\Omega \times [0,\delta] \times [0,\bar{\varepsilon}]\,)$ such that

$$\lambda_n(\hbar,\varepsilon) = f^{\infty}(n\hbar,\varepsilon) + \hbar g(n\hbar,\hbar,\varepsilon) + \hbar|\omega|/2 + O(\varepsilon^{\infty}) \tag{1.29}$$

whenever $n\hbar \in \Gamma^{\infty}(\varepsilon)$. Here $\lambda_n(\hbar,\varepsilon)$ is any eigenvalue of $T(\hbar,\varepsilon)$ close to $\lambda_n(\hbar)$ for $\varepsilon < \bar{\varepsilon}$, and $O(\varepsilon^{\infty})$ is uniform with respect to $\hbar \in [0,\delta]$.

Remark.

(1.29) is the natural generalization of the Bohr-Sommerfeld quantization formula, to all orders in perturbation theory, for the present class of non-integrable systems.

2. Proof of the main results.

Consider first Proposition 1. Assertion (1) is a direct consequence of the following formulae

$$P_o(\hbar,\omega)\psi/\psi = \langle \omega z, \nabla_z \rangle W(z,\cdot) - \omega z^2 \tag{2.1}$$

$$V(z+\hbar\nabla_z)/\sqrt{2\omega})e^{\{W(z,\cdot)-z^2/2\}/\hbar} = e^{-z^2/2\hbar} V(\hbar\nabla_z/\sqrt{2\omega})e^{W(\cdot)/\hbar} \tag{2.2}$$

$$e^{-W(\cdot)/\hbar} V(\hbar\nabla_z/\sqrt{2\omega})e^{W(\cdot)/\hbar} = V(\nabla_z W(\cdot)/\sqrt{2\omega}) + \sum_{\ell=1}^{2m} \hbar^{\ell} R_{\ell}(W(\cdot)) \tag{2.3}$$

Formula (2.1) is immediate; (2.2) is proved e.g. in Voros[6]; the lengthy combinatorical proof of (2.3) is given in detail in Ref.(1) and is therefore omitted here.

Consider now Assertion (2). Its "classical" part can be verified by direct inspection. To verify the remaining part, remark that equation (1.18) for $\varepsilon = 0$ follows by the Schrödinger equation $P_o(\hbar,\omega)\psi(z,E) = E\psi(z,E)$ upon insertion of

$$\psi(z,E) = e^{\{W_o(z,E)-z^2/2\}/\hbar} \tag{2.4}$$

Look now for solutions of (1.18), rewritten here as

$$< \omega z, \nabla_z > W_o(z,E) - \omega z^2 = E \tag{2.5}$$

in separated variables:

$$E = \sum_{i=1}^{d} E_i,\ W_o(z,E) = \sum_{i=1}^{d} W_i^o(z_i,E_i) \tag{2.6}$$

Then (2.5) yields

$$\partial_i W_o^i(z_i,E_i) = (E_i + \omega_i z_i^2)/\omega_i z_i \quad , \ i = 1,\dots,d. \tag{2.7}$$

Since $z \to \psi(z,E)$ is an eigenfunction only if it is holomorphic in $\mathbb{C}^d$, the quantization condition is obtained:

$$\int_\Gamma \frac{\partial_i \psi_i(z_i,E_i)}{\psi_i(z_i,E_i)} dz_i = \hbar^{-1} \int_\Gamma \partial_i W_o^i(z_i,E_i) dz_i = n_i,\ n_i = 0,1,\dots \tag{2.8}$$

Γ being any circumference in $\mathbb{C}$ avoiding the zeros of $\psi_i(z_i,E_i) = \exp\{(W_o^i(z_i,E_i) - z^2/2)/\hbar\}$. By (2.7) it is $\hbar^{-1} \int_\Gamma \partial_i W_o^i(z_i,E_i) dz_i = E_i/\omega_i$. Therefore:

$$E_i = \hbar n_i \omega_i,\ E = \hbar < n,\omega > = \lambda_n(\hbar) - \hbar|\omega|/2 = f_o(n\hbar,\omega) \tag{2.9}$$

$$W_o^i(z_i,E_i) = \hbar \log(z_i)^{n_i} + z_i^2/2,\ \psi(z,E) = C_{n_1\dots n_d} z^{n_1} .. z^{n_d} \tag{2.10}$$

which are of course the familiar eigenfunctions (in the Bargmann representation) and eigenvalues of the quantum harmonic oscillator. $C_{n_1..n_d}$ are the usual normalization constants.

(3) Following the well known Birkhoff procedure of classical, canonical perturbation theory (see e.g. Gallavotti [3], § 5.10) set, heuristically:

$$W(A,z,\hbar,\varepsilon) = \sum_{k=o}^{\infty} W_k(A,z,\hbar)\varepsilon^k \tag{2.11}$$

$$E(A,\hbar,\varepsilon) = \sum_{k=o}^{\infty} \lambda_k(A,\hbar)\varepsilon^k \tag{2.12}$$

where W_o, λ_o are given by (1.19) and (1.3), respectively.

Insertion in (1.18), Taylor expansion of both sides near $\varepsilon = 0$ and equality of k-th order coefficients yields:

$$\begin{aligned} &< \omega z, \nabla_z > W_k(A,z,\hbar) + \frac{1}{(k-1)!} \frac{d^{k-1}}{d\varepsilon^{k-1}} \{ V(\sum_{j=0}^{k-1} \varepsilon^j \nabla_z W_j(\cdot)/\sqrt{2\omega}) + \\ &+ \sum_{\ell=1}^{2m} \hbar^\ell R_\ell (\sum_{j=o}^{k-1} \varepsilon^j W_j(\cdot))\}\Big|_{\varepsilon=o} = \lambda_k(A,\hbar),\ k = 1,\dots \end{aligned} \tag{2.13}$$

It is seen at once that the second and the third addend in the ℓ.h.s. of (2.13) depend on $(W_o,\dots,W_j)$ and on their derivatives only up to $j = k-1$.

Set therefore:

$$\frac{1}{(k-1)}\frac{d^{k-1}}{d\varepsilon^{k-1}} V\Big(\sum_{j=o}^{k-1}\varepsilon^j\nabla_z W_j(\cdot)/\sqrt{2\omega}\Big)\Big|_{\varepsilon=o} = Y_k(W_o,\dots,W_{k-1}) \tag{2.14}$$

$$\frac{1}{(k-1)!}\frac{d^{k-1}}{d\varepsilon^{k-1}} R_\ell\Big(\sum_{j=1}^{k-1}\varepsilon^j W_j(\cdot)\Big)\Big|_{\varepsilon=o} = Z_k^\ell(W_o,\dots,W_{k-1}) \tag{2.15}$$

The explicit expressions of $Y_k(\cdot)$, $Z_k^\ell(\cdot)$ can be found in Ref.(1). Rewrite now (2.13) in abbreviated form:

$$<\omega z,\nabla_z> W_k(A,\text{\textcrh},z) + Y_k(W_o,\dots,W_{k-1}) + \sum_{\ell=1}^{2m}\text{\textcrh}^\ell Z_k^\ell(W_o,\dots,W_{k-1}) = \lambda_k(A,\text{\textcrh}),\quad k = 1,\dots. \tag{2.16}$$

Since W_o and λ_o are known, eq. (2.16) becomes a recursive, inhomogeneous, linear first order equation for W_k and λ_k. To find W_k and λ_k, look for the Laurent expansion of $W_k(A,z,\text{\textcrh})$:

$$W_k(A,z,\text{\textcrh}) = \sum_{q\in\mathbb{Z}^d} w_q^{(k)}(A,\text{\textcrh})z^q,\quad z^q = z_1^{q_1}\dots z_d^{q_d} \tag{2.17}$$

Inserting this in (2.16) one has:

$$<\omega,q> w_q^{(k)}(A,\text{\textcrh}) + g_q^{(k)}(A,\text{\textcrh}) + \sum_{\ell=1}^{2km}\text{\textcrh}^\ell\theta_q^{(k),\ell}(A) = 0,\quad q\neq 0 \tag{2.18}$$

$$y_o^{(k)}(A,\text{\textcrh}) + \sum_{\ell=1}^{2km}\text{\textcrh}^\ell\theta_o^{(k),\ell}(A) = \lambda_k(n,\text{\textcrh}),\quad q = 0 \tag{2.19}$$

if the identity of the Laurent expansions of both sides is required. Here $y_q^{(k)}(A)$, $\theta_q^{(k),\ell}(A)$ stand for the Laurent coefficients of $Y_k(W_o,\dots,W_{k-1})$, $Z_k^\ell(W_o,\dots,W_{k-1})$, respectively. The solution of (2.18) is of course:

$$w_q^{(k)}(A,\text{\textcrh}) = -\Big[y_q^{(k)}(A,\text{\textcrh}) + \sum_{\ell=1}^{2km}\text{\textcrh}^\ell\theta_q^{(k),\ell}(A)\Big]/<\omega,q> \tag{2.20}$$

Consider first the case $k = 1$. By the explicitt expression of $W_o(A,z)$ it is immediately seen that $A\to y_q^{(1)}(A)$, $A\to\theta_q^{(1),\ell}(A)$ are holomorphic in Ω, so that $A\to w_q^{(1)}(A)$ has the same property. Furthermore, since

V is a polynomial, by (1.2) there are $C_1>0$, $C_2>0$, $D>0$ such that, for any $\alpha>0$:

$$\sup_{A\in\Omega} |y_q^{(1)}(A)/<\omega,q>| \le C_1 e^{-\alpha|q|} \tag{2.21}$$

$$\sup_{A\in\Omega} | \theta_q^{(1),\ell}(A)/<\omega,q>| \le C_2\cdot D^{\ell} e^{-\alpha|q|} \tag{2.22}$$

By (2.17) and (2.20) is exhibited as the sum of a Laurent series convergent in $(\mathbb{C}\setminus\{0\})^d$, uniformly with respect to $(A,\not{h})\in\Omega\times\mathbb{C}$. Therefore it has the stated holomorphy properties and the form (1.21). Furthermore the following explicit expressions are obtained for $P_1(A)$ and $Q_1^{\ell}(A)$:

$$P_1(A) = y_o^{(1)}(A),\quad Q_1^{\ell}(A) = \theta_o^{(1),\ell}(A) \tag{2.23}$$

Since the Laurent coefficients of $W_1(A,\not{h})$ vanish faster than any negative exponential, the argument can immediately be iterated to all $k\ge 2$, and this proves Assertion (3).

(4) The assertion is true for $k=0$ as it has been seen above. Consider now the classical Hamiltonian $H(\omega,\varepsilon)$ and reexpress it in the (R,z) canonical variables given by (1.6):

$$H(C_o^{-1}(R,z);\varepsilon) = H_2(R,z;\varepsilon) = F_o(z,R) + \varepsilon V(R/\sqrt{2}\omega) \tag{2.24}$$

Now, once more according to the Birkhoff transformation of canonical perturbation theory, look for a completely canonical bijection $C_\varepsilon(R,z) = (A,\phi)$ of $(\mathbb{C}\setminus\{0\})^d \cong \mathbb{R}_+^d\times\mathbb{T}^d$ such that $H_2(C_\varepsilon^{-1}(A,\phi))$ has a formal expansion in powers of ε with coefficients depending only on A. To this end, look for the generating function $\Phi(A,z;\varepsilon)$ of C_ε :

$$\begin{cases} R = \nabla_z\Phi(A,z,\varepsilon) \\ \phi = i\nabla_A\Phi(A,z,\varepsilon) \end{cases} \tag{2.25}$$

under the form of a formal power series in ε :

$$\Phi(A,z;\varepsilon) = \sum_{k=o}^{\infty} \Phi_k(A,z)\varepsilon^k \tag{2.26}$$

where $\Phi_k(\cdot)$, $k \geq 1$, are to be recursively determined out of the Hamilton-Jacobi equation since $\Phi_o(A,z) = W_o(A,z)$ is known by assertion (2) above. Look once more for the Laurent coefficients of $\Phi_k(A,z)$:

$$\Phi_k(A,z) = \sum_{q \in \mathbb{Z}^d} \Phi_q^{(k)}(A) z^q \tag{2.27}$$

Then, once more upon insertion of (2.26), (2.27) in the Hamilton-Jacobi equation:

$$\begin{aligned} F_o(z, \nabla_z \Phi(A,z,\varepsilon)) + \varepsilon V(\nabla_z \Phi(A,z,\varepsilon)/\sqrt{2\omega}) = \\ = \text{function depending only on } (A,\varepsilon) \end{aligned} \tag{2.28}$$

and a universal expansion in powers of ε, the request that the coefficients of the r.h.s. be z-independent yields:

$$< \omega, q > \Phi_q^{(k)}(A) + Y_q^{(k)}(A) = 0, \; k \geq 1, \; q \neq 0 \tag{2.29}$$

$$Y_k(A) = Y_o^{(k)}(A) \tag{2.30}$$

where $A \to Y_q^{(k)}(A)$ are the Laurent coefficient of $Y_k(W_o,\ldots,W_{k-1}) = Y_k(\Phi_o,\ldots,\Phi_{k-1})$ defined by (2.14). Therefore the recursive equations (2.29), (2.30) are identical to the recursive equations (2.18), (2.19) with $\hbar = 0$ and have the same initial conditions because $\Phi_o = W_o$. This proves that $A \to P_k(A)$ is the k-th term of the Birkhoff expansion. Furthemore, when $\hbar \neq 0$ the algorithm yields a formal power series expansion for each eigenvalue $\lambda_n(\hbar,\varepsilon)$ of $T(\omega,\varepsilon)$, if we make $A = n\hbar$. Since under the present assumptions $\lambda_n(\hbar,\varepsilon)$ admits the Rayleigh-Schrödinger series as an asymptotic expansion to all orders in ε, and a function can have at most one asymptotic expansion, the coefficients $\lambda_k(n\hbar, \hbar)$ coincide with the Rayleigh-Schrödinger ones. This concludes the proof of Proposition 1.

Given the KAM theorem and the C^∞ version of the Borel theorem (see e.g.

Hörmander [(6)]) Corollary 1 is essentially a rephrasing of the fact that the perturbation expansion exist to all order and that its divergence can be controlled uniformly with respect to $\hbar$ in the limit $n\hbar \to A$.

Proof of Corollary 1.

Set, for any $\delta > 0$:

$$\beta_k = \sup_{A\in\Omega, 0\le\hbar\le\delta} \sum_{\ell=0}^{2mk-1} Q_k^{\ell+1}(A)\hbar^\ell \quad , \; k = 1,2,.... \tag{2.31}$$

$$F_k(A,\hbar) = \sum_{\ell=0}^{2mk-1} Q_k^{\ell+1}(A)\hbar^\ell \quad , \; k = 1,2,.... \tag{2.32}$$

Therefore $(A,\hbar) \to F_k(A,\hbar) \in C^\infty(\Omega \times [0,\delta])$, with $\sup_{A\,\Omega, 0\le\hbar\le\delta} |F_k(A,\hbar)| \le \beta_k$. Let $x \to \chi(x) \in C_0^\infty(\mathbb{R})$, $\chi(x) = 1$ for $|x| \le 1$, $\chi(x) = 0$ for $|x| > 2$, and let $\{\gamma_k\}_{k=0}^\infty$ be a positive sequence increasing monotonically to $+\infty$. Set:

$$g_N(A,\hbar,\varepsilon) = \sum_{k=0}^{N-1} F_k(A,\hbar)\,\chi(\varepsilon\gamma_k), \; N = 1,2,... \tag{2.33}$$

By [6, Theorem 1.2.6] it is possible to choose the sequence $\{\gamma_n\}$, depending on $\{\beta_k\}$ and thus not on $\hbar$, such that there is $(A,\hbar,\varepsilon) \to$ $\to g^\infty(A,\hbar,\varepsilon) \in C^\infty(\Omega \times [0,\delta] \times [0,\bar{\varepsilon}])$, $\bar{\varepsilon}$ suitably small, with the property:

$$g^\infty(A,\hbar,\varepsilon) - \sum_{k=1}^{N-1} F_k(A,\hbar)\varepsilon^k = O(\varepsilon^N) \quad , \; N = 1,2,... \tag{2.34}$$

On the other hand, if $A = n\hbar$ is a KAM torus, i.e. if $n\hbar \in \Gamma^\infty(\varepsilon)$, one has:

$$f^\infty(n\hbar,\varepsilon) - \sum_{k=0}^{N-1} P_k(n\hbar)\varepsilon^k = O(\varepsilon^N), \quad N = 1,... \tag{2.35}$$

Since under the present assumptions on V it is well known that, for any fixed $\hbar > 0$, one has:

$$\lambda_n(\hbar,\varepsilon) - \sum_{k=0}^{N-1} \lambda_n^k(\hbar)\varepsilon^k = (\varepsilon^N), \quad N = 1,... \tag{2.36}$$

the assertion is proved because $\lambda_n^k(\hbar) = P_k(n\hbar) + \hbar F_k(n\hbar,\hbar)$.

References

(1) S.Graffi and T.Paul: Schrödinger Equation and Canonical Perturbation Theory. Submitted to Commun.Math.Phys.

(2) V.Bargmann: On a Hilbert Space of Analytic Functions and an Associated Integral Transform, I. Commun.Pure Appl.Math.14, 187-214 (1961).

(3) G.Gallavotti: Elementary Mechanics, Springer-Verlag 1983.

(4) L.Chierchia and G.Gallavotti: Snooth Prime Integrals for Quasi-Integrable Hamiltonian Systems. Nuovo Cimento 67 B , 277-295 (1982).

(5) G.Gallavotti: Perturbation Theory for Classical Hamiltonian Systems. In: Progress in Physics, J.Fröhlich, Editor, Birkhäuser 1982.

(6) L.Hörmander: The Analysis of Linear Partial Differential Operators, I. Springer-Verlag 1983.

TRACE ESTIMATES FOR EXTERIOR BOUNDARY PROBLEMS ASSOCIATED WITH THE SCHRÖDINGER OPERATOR

Gerd Grubb
Mathematics Department, Copenhagen University
Universitetsparken 5, DK-2100 Copenhagen, Denmark

The present work is concerned with some mathematical questions arizing in connection with the study of the wave equation in a domain exterior to a bounded obstacle, in acoustic scattering theory. We study the operator defined as the difference between the free space resolvent and the resolvent with a boundary condition, and we use this to study similar differences between functions of the free space operator and the operator with a boundary condition, typically functions such as $\lambda \sim \exp(-t\lambda)$ and $\lambda \sim \lambda^z$. It is shown that each difference operator is the sum of a wellknown contribution from the interior of the obstacle (giving the principal part of the spectral behavior in the same ways as operators on bounded domains) plus a contribution that "lives" near the boundary of the obstacle; we analyze the boundary part in detail, and show how the space dimension n is replaced by the boundary dimension $n-1$ in spectral estimates. (Dimensional observations of this kind go back to Birman [4], who studied the difference between the Dirichlet and Neumann problem solution operators in an exterior domain.)

The resolvent difference operator is studied within the framework of the Boutet de Monvel theory [6], [12], whereby one obtains a detailed description of its structure as a pseudo-differential boundary operator. For the difference operator connected with $\exp(-t\lambda)$ (the "heat operator" difference) we obtain as a consequence a full asymptotic expansion of the trace, completing earlier, partial results of Deift and Simon [7], Jensen and Kato [14], Majda and Ralston [17]. For the fractional power difference we get spectral asymptotic estimates, characterizing in particular the exponents for which one has trace class operators; this improves and generalizes observations of Deift and Simon [7], Reed and Simon [19], Bardos, Guillot and Ralston [3]. - It is our hope that this clarification of the structure of some tools in the theory may also lead to an improved understanding of the main questions.

The results presented below partly rephrase, partly extend the results of Grubb [11]: the order of the elliptic operator is now allowed to be $2m$ instead of 2 in [11], and the spectral estimates are made more precise, in particular for the fractional powers. We give a full account of the extensions, referring to [11] for the arguments that are covered there.

1. Introduction

Let Ω_- be a bounded open set in $\mathbb{R}^n$ (the obstacle) with smooth boundary, and let $\Omega_+ = \mathbb{R}^n \setminus \overline{\Omega}_-$, the exterior of Ω_-. The two sets have the common boundary $\Gamma = \partial\Omega_+ = \partial\Omega_-$, which is a smooth compact $(n-1)$-dimensional C^∞ manifold.

Let A be a $2m$-order formally selfadjoint, uniformly strongly elliptic differential operator on $\mathbb{R}^n$ with C^∞ coefficients, defining a bijection of $H^{s+2m}(\mathbb{R}^n)$ onto $H^s(\mathbb{R}^n)$ for any s, and satisfying (with positive constants c_1 and c_2)

$$(1) \qquad c_1\|u\|_m^2 \leq (Au,u) \leq c_2\|u\|_m^2$$

for $u \in H^{2m}(\mathbb{R}^n)$. (Here $H^k(\Omega)$ denotes for integer $k \geq 0$ the Sobolev space of L^2 functions on Ω with L^2 derivatives up to order k and with norm $\|u\|_k = (\Sigma_{|\alpha|\leq k}\|D^\alpha u\|_{L^2}^2)^{\frac{1}{2}}$. The generalizations to noninteger and negative k are described e.g. in Lions-Magenes [16].)

Let A_+ resp. A_- denote closed realizations of A on Ω_+ resp. Ω_-, defined by smooth normal boundary conditions at Γ (for example Dirichlet or Neumann conditions). We assume that these boundary conditions satisfy the Shapiro-Lopatinski conditions (define elliptic realizations), and that the inequality (1) holds for $u \in D(A_\pm)$; for the discussion of (1), cf. Agmon [1], Grubb [9].

Example. For A, one can take e.g. $-\Delta+V(x)$ (with $0 \leq c_1' \leq V(x) \leq c_2'$) or, more generally, $A = \Sigma_{j,k=1}^n a_{jk}(x)D_jD_k + \Sigma_{j=1}^n a_j(x)D_j + a_0(x)$, with $\Sigma_{j,k=1}^n a_{jk}(x)\xi_j\xi_k$ positive definite (for $\xi\in\mathbb{R}^n$) and with suitable bounds on the coefficients. (Here $D_j = -i\partial/\partial x_j$.) In this case, $m = 1$. As realizations A_+ and A_-, one can then take the operator acting like A and with domain determined by the Dirichlet or a Neumann or Robin condition (assuming that the boundary problem is associated with a positive sesquilinear form, as described in [11]).

Then also A^N , $(A_+)^N$ and $(A_-)^N$ (N integer ≥ 1) satisfy our hypotheses (the operators $(A_\pm)^N$ are realizations of A^N).

In acoustic scattering theory, one compares the solutions of the free wave equation (for A) with the solutions of the exterior wave equation (for A_+) . One of the tools here is the study of the <u>resolvent differences</u>

(2) $$B_\lambda = (A-\lambda)^{-1} - (A_+-\lambda)^{-1} \oplus 0$$

(3) $$G_\lambda = (A-\lambda)^{-1} - (A_+-\lambda)^{-1} \oplus (A_--\lambda)^{-1} .$$

(We denote by $S_+ \oplus S_-$ the sum of the operators S_+ on $L^2(\Omega_+)$ resp. S_- on $L^2(\Omega_-)$, extended by zero on $L^2(\Omega_-)$ resp. $L^2(\Omega_+)$; here $L^2(\Omega_\pm)$ are identified naturally with subspaces of $L^2(\mathbb{R}^n)$.) More generally, one uses the <u>operator function differences</u>

(4) $$B_f = f(A) - f(A_+) \oplus 0$$

(5) $$G_f = f(A) - f(A_+) \oplus f(A_-) ;$$

here we define $f(P)$ by means of a Cauchy integral formula

$$f(P) = \frac{i}{2\pi} \int_L f(\lambda)(P-\lambda)^{-1} d\lambda ,$$

where f is holomorphic on a neighborhood V of the spectrum of P in $\mathbb{C}$, L is a curve in V going around the spectrum in the positive direction, and the integral converges in a suitable sense. Note that

(6) $$B_\lambda = G_\lambda + 0 \oplus (A_--\lambda)^{-1} ;$$

(7) $$B_f = G_f + 0 \oplus f(A_-) .$$

As we shall see, G_λ and G_f are in a sense finer than B_λ and B_f , which contain the "solid" contribution from Ω_- .

As examples of such operator functions, we consider the cases $f(\lambda) = \exp(-t\lambda)$ $(t > 0)$ and $f(\lambda) = \lambda^z$ (Re $z < 0$). The results for the resolvent differences (2) and (3) have the form of very thorough informations on their symbolic structure (in the framework of pseudo-differential boundary problems) and spectral properties, for λ on rays in $\mathbb{C}$ disjoint from $\mathbb{R}_+$ - with special consequences for the traces (in case $2m > n$, or when A is replaced by A^N , $2mN > n$). Before discussing these resolvent results, we describe their consequences for the abovementioned operator functions.

We use the notation $\lambda_k^{\pm}(T)$ for the positive resp. negative eigenvalues of a selfadjoint compact operator, arranged in a nonincreasing, resp. nondecreasing sequence (repeated according to multiplicities). More generally, when T is a compact operator in a Hilbert space H, the characteristic values $s_k(T)$ are defined as the positive eigenvalues $\lambda_k^+(|T|)$ of $|T| = (T^*T)^{\frac{1}{2}}$. We denote by $\mathfrak{C}_p$ the Schatten class (trace ideal) of compact operators T in H with

$$\|T\|_{\mathfrak{C}_p} \equiv \left(\sum_{k=1}^{\infty} s_k(T)^p\right)^{1/p} < \infty ,$$

in particular, $\mathfrak{C}_1$ and $\mathfrak{C}_2$ consist of the trace class resp. Hilbert-Schmidt operators. We shall also use the space $\mathfrak{S}_{(p)}$ of compact operators T in H with

$$N_p(T) \equiv \sup_k s_k(T)k^{1/p} < \infty ; \tag{8}$$

here $N_p(T)$ is a quasi-norm satisfying e.g.

$$N_p\left(\sum_{\ell=1}^{M} T_\ell\right) \le C_{p,r} \sum_{\ell=1}^{M} N_p(T_\ell)\ell^r \quad \text{for } r > p^{-1} - 1 , \text{ if } p \le 1 , \tag{9}$$

cf. Birman-Solomiak [5]. The results concerning this quasi-norm are often sharper than those concerned with the $\mathfrak{C}_p$-norms. Note that $\mathfrak{S}_{(p)} \subset \mathfrak{C}_{p+\varepsilon}$ for $\varepsilon > 0$.

The results are valid also for matrixformed operators (and operators in vector bundles), by immediate generalizations of the proofs.

2. The heat operator difference.

We here consider

$$(10) \qquad \begin{aligned} B_{\exp}(-t) &= \exp(-tA) - \exp(-tA_+) \oplus 0 , \\ G_{\exp}(-t) &= \exp(-tA) - \exp(-tA_+) \oplus \exp(-tA_-) , \end{aligned}$$

and observe the obvious relation

$$B_{\exp}(-t) = G_{\exp}(-t) + 0 \oplus \exp(-tA_-) . \tag{11}$$

For the case $A = -\Delta$, with the realizations A_+ and A_- determined by the Dirichlet condition at Γ, Deift and Simon [7] observed that $G_{\exp}(-t)$ is of trace class for $t > 0$. In view of the wellknown fact that $\exp(-tA_-)$ is of trace class for each $t > 0$ (in fact the eigenvalues go to zero rapidly), $B_{\exp}(-t)$ is likewise of trace class.

Jensen and Kato [14] showed the asymptotic estimate for the trace

(12) $$\operatorname{tr} B_{\exp}(-t) = b_{-n}t^{-n/2} + O(t^{(1-n)/2}) \quad \text{for} \quad t \to 0+ ,$$

where b_{-n} is the <u>same</u> positive constant as in the wellknown analogous estimate for $\operatorname{tr}\exp(-tA_-)$. They also showed estimates for trace <u>norms</u>, in particular

$$\|G_{\exp}(-t)\|_{\mathfrak{C}_1} = O(t^{(1-n)/2}) \quad \text{for} \quad t \to 0+ .$$

(These authors require somewhat less moothness of Γ than we do.) Majda and Ralston [17] extended the result (12) to give three terms plus a remainder

$$\operatorname{tr} B_{\exp}(-t) = \sum_{j=0,1,2} b_{j-n}t^{(j-n)/2} + O(t^{(3-n)/2}) \quad \text{for} \quad t \to 0+$$

by use of the expansion on bounded domains (also with three precise terms) of McKean and Singer [18].

Our analysis leads to the following result, under the assumptions we have described:

<u>Theorem 1</u>.

1° $G_{\exp}(-t)$ *belongs to* $\bigcap_{p>0} \mathfrak{S}_{(p)}$ *for each* $t > 0$. *The trace has the asymptotic expansion*

(13) $$\operatorname{tr} G_{\exp}(-t) \sim \sum_{j=1}^{\infty} e_{j-n}t^{(j-n)/2m} \quad \textit{for} \quad t \to 0+ ,$$

where e_{j-n} *is determined from the* $j+1$ *top order symbols of* A *and the* j *top order symbols of the boundary conditions. Moreover, the trace norm satisfies*

(14) $$\|G_{\exp}(-t)\|_{\mathfrak{C}_1} = O(t^{(1-n)/2m}) \quad \textit{for} \quad t \to 0+ .$$

2° $B_{\exp}(-t)$ *belongs to* $\bigcap_{p>0} \mathfrak{S}_{(p)}$ *for each* $t > 0$. *The trace has the asymptotic expansion*

(15) $$\operatorname{tr} B_{\exp}(-t) \sim \sum_{j=0}^{\infty} b_{j-n}t^{(j-n)/2m} \quad \textit{for} \quad t \to 0+ ,$$

with

(16) $$\|B_{\exp}(-t)\|_{\mathfrak{C}_1} = b_{-n}t^{-n/2m} + O(t^{(1-n)/2m}) \quad \textit{for} \quad t \to 0+ .$$

Here, b_{-n} *is identical with the leading constant in the asymptotic expansion of* $\mathrm{tr}\exp(-tA_-)$ *(cf. Seeley [22], Greiner [8]). In fact, writing*

$$\mathrm{tr}\exp(-tA_-) \sim \sum_{j=0}^{\infty} a_{j-n} t^{(j-n)/2m} \quad \text{for} \quad t \to 0+ \,, \tag{17}$$

we have that

$$\begin{aligned} b_{-n} &= a_{-n} \,; \\ b_{j-n} &= a_{j-n} + e_{j-n} \quad \text{for} \quad j > 0 \,. \end{aligned} \tag{18}$$

Observe in particular the fact, that the asymptotic series for $\mathrm{tr}\, G_{\exp}(-t)$ begins with $j = 1$ whereas the series for $\mathrm{tr}\, B_{\exp}(-t)$ begins with $j = 0$, so that they add up as in (18) (by use of (11)). This shows why the series for $\mathrm{tr}\, B_{\exp}(-t)$ and for $\mathrm{tr}\exp(-tA_-)$ have the same leading term; this phenomenon has been expressed in the literature as the "Weyl formula" for $B_{\exp}(-t)$ (leading to "Weyl formulas" for the scattering phase and the spectral shift function, cf. Jensen-Kato [14] and Majda-Ralston [17]).

3. Fractional powers.

One can also consider

$$\begin{aligned} B^{(-s)} &= A^{-s} - (A_+)^{-s} \oplus 0 \,, \\ G^{(-s)} &= A^{-s} - (A_+)^{-s} \oplus (A_-)^{-s} \,, \quad \text{for} \quad s > 0 \,. \end{aligned} \tag{19}$$

For the case where A equals $-\Delta+V(x)+1$ for some nonnegative $V \in L^1$, Ω_- is a ball, and $A_\pm$ are defined by Dirichlet boundary conditions, Deift and Simon [7] showed that $G^{(-s)}$ is of trace class if $s \geq 2$, $s > n/2$. Reed and Simon [19] included the case of Neumann boundary conditions, when $A = -\Delta+1$ and $s = 2$, $n = 3$ (and Ω_- is a finite union of starshaped domains). Bardos, Guillot and Ralston [3] sharpened the result for $-\Delta+1$ in the Dirichlet case, by showing how it follows from the estimates of Jensen and Kato [14] that $B^{(-s)}$ is of trace class for $s > n/2$. This implies that $G^{(-s)}$ is likewise of trace class when $s > n/2$, since we have that

$$B^{(-s)} = G^{(-s)} + 0 \oplus (A_-)^{-s} \,, \tag{20}$$

where the fact that $(A_-)^{-s}$ is of trace class for $s > n/2$ is an immediate consequence of wellknown eigenvalue estimates for A_-.

However (as Deift and Simon also remark), $G^{(-s)}$ should have a finer behavior than $B^{(-s)}$. This is indeed the case, and is contained in our general result.

Theorem 2. *Let* $s > 0$.

1° $G^{(-s)}$ *is in the space* $\mathfrak{S}_{(p)}$ *with* $p = (n-1)/2ms$. *In other words, there are constants* $c_{\pm}$ *so that*

$$(21)\qquad |\lambda_k^{\pm}(G^{(-s)})| \leq c_{\pm}k^{-2ms/(n-1)} \quad \textit{for} \quad k = 1,2,\ldots .$$

In particular, $G^{(-s)}$ *is of trace class, when* $s > (n-1)/2m$.

2° *The eigenvalues of* $B^{(-s)}$ *satisfy*

$$(22)\qquad \lambda_k^{+}(B^{(-s)}) = C_s k^{-2ms/n} + O(k^{-(2ms+\delta)/n}) \quad \textit{for} \quad k \to \infty,$$

$$(23)\qquad |\lambda_k^{-}(B^{(-s)})| = O(k^{-2ms/(n-1)}) \quad \textit{for} \quad k \to \infty,$$

where $\delta = 2ms/(2ms+n-1)$. *Here* C_s *is the same (positive) constant as in the wellknown estimate* $\lambda_k(A_-^{(-s)}) \sim C_s k^{-2ms/n}$. *In particular,* $B^{(-s)}$ *is of trace class when* $s > n/2m$.

Remark. The estimate (22) is equivalent with the estimate of the "counting function"

$$(24)\qquad N^{+}(t;B^{(-s)}) = C_s' t^{n/2ms} + O(t^{(n-\delta)/2ms}) \quad \text{for} \quad t \to \infty,$$

where $N^{+}(t;B^{(-s)})$ stands for the number of eigenvalues larger than $1/t$, and $C_s' = C_s^{-n/2ms}$ (cf. e.g. Grubb [10, Lemma 6.2]).

In the cases where s is an integer, there are sharper remainder estimates than (22), namely, δ can be any real number less than 1, cf. Grubb [12, Th. 5.3].

The above estimates for real s extend to the non-selfadjoint operators $G^{(z)}$ and $B^{(z)}$ constructed from complex powers A^z, $A_{\pm}^z$, for $\text{Re } z < 0$, in the following way:

$$(25)\qquad \begin{aligned} s_k(G^{(z)}) &\leq Ck^{-2m|\text{Re } z|/(n-1)}, \\ s_k(B^{(z)}) &= C_z k^{-2m|\text{Re } z|/n} + O(k^{-(2m|\text{Re } z|+\delta)/n}) \quad \text{for } k \to \infty, \end{aligned}$$

with $\delta = 2m|\text{Re } z|/(2m|\text{Re } z|+n-1)$.

Observe the general principle in Theorem 2 as well as Theorem 1: in the contributions from the G-terms, the usual dimension n is replaced by the boundary dimension $n-1$.

4. The proofs, and the underlying results for the resolvent.

The proofs have basically two ingredients: one is the elimination of the contributions from the part of $\mathbb{R}^n$ far from the obstacle, and the other is a study of the structure of the operators (especially the G-terms) near Γ .

The starting point is a global estimate of the resolvent.

Lemma 1. *For each argument* $\theta \in]0,2\pi[$ *there is a constant* $c_\theta > 0$ *such that for all* $\lambda = re^{i\theta}$ $(r \geq 0)$,

$$(26) \qquad \|(A_\pm-\lambda)^{-1}f\|_{2m} + (1+|\lambda|)\|(A_\pm-\lambda)^{-1}f\|_0 \leq c_\theta\|f\|_0 \quad \textit{for} \quad f \in L^2(\Omega_\pm) .$$

Similar estimates hold for $(A-\lambda)^{-1}$.

Proof: In the case $m = 1$, this is Lemma 3.4 in [11]. For completeness' sake, we now show how it is obtained for all $m \geq 1$. Consider $(A_+-\lambda)^{-1}$, and fix θ . By the positivity of A_+ , one finds

$$(27) \qquad (1+|\lambda|)\|(A_+-\lambda)^{-1}f\|_0 \leq c_1\|f\|_0$$

as in [11, (3.13)]. Now let $\chi \in C_0^\infty(\mathbb{R}^n)$, $\chi = 1$ on a neighborhood of $\overline{\Omega}_-$. By the ellipticity of the boundary value problem, one has for $u \in D(A_+)$ that $\chi u \in H^{2m}(\Omega_+)$ and hence $A\chi u \in L^2(\Omega_+)$. Then also $A(1-\chi)u = Au-A\chi u \in L^2(\Omega_+)$, which implies $(1-\chi)u \in H^{2m}(\mathbb{R}^n)$ by the bijectiveness property of A . Thus altogether, $u = \chi u + (1-\chi)u$ is in $H^{2m}(\Omega_+)$, so we have $D(A_+) \subset H^{2m}(\Omega_+)$, and the inequality

$$\|u\|_{H^{2m}(\Omega_+)} \leq c_2(\|u\|_{L^2(\Omega_+)} + \|Au\|_{L^2(\Omega_+)})$$

is valid for $u \in D(A_+)$. In particular, when $u = (A_+-\lambda)^{-1}f$ for some $f \in L^2(\Omega_+)$, this implies

$$\|(A_+-\lambda)^{-1}f\|_{2m} \leq c_2(\|(A_+-\lambda)^{-1}f\|_0 + \|f+\lambda(A_+-\lambda)^{-1}f\|_0) \leq c_3\|f\|_0$$

in view of (27). This gives (26) for A_+ ; the proofs for A_- and

A are similar or easier.

The estimate (26) implies by interpolation

$$\|(A_\pm-\lambda)^{-1}f\|_{2m-s} \leq c'_\theta(1+|\lambda|)^{-s/2m}\|f\|_0\ , \quad f \in L^2(\Omega_\pm)\ ,$$

for any real $s \in [0,2m]$. Now the commutator $[A,\chi]$ (with χ as defined in the above proof) is a differential operator of order $2m-1$, so (for fixed θ)

$$\|[A,\chi](A_+-\lambda)^{-1}f\|_{\frac{1}{2}} \leq c_1\|(A_+-\lambda)^{-1}f\|_{2m-\frac{1}{2}} \leq c_2(1+|\lambda|)^{-1/4m}\|f\|_0\ .$$

Since $[A,\chi]$ ranges in a space of functions supported in $\operatorname{supp}\chi$, it follows from wellknown spectral estimates connected with Sobolev imbeddings that the operator $[A,\chi](A_+-\lambda)^{-1}$ in $L^2(\Omega_+)$ is compact, with characteristic values satisfying

$$(28)\qquad s_k([A,\chi](A_+-\lambda)^{-1}) \leq c_3(1+|\lambda|)^{-1/4m}k^{-1/2n} \quad \text{for } k=1,2,\ldots\ .$$

There are similar estimates with A_+ replaced by A_- or A.

We use this in a treatment of the contribution from G_λ (cf. (3)) near Γ , where the main trick is to introduce a factorization containing a large number of such commutators with cut-off functions, by use of the formula

$$\begin{aligned}\chi_1[(A-\lambda)^{-1},\chi]u &= \chi_1[(A-\lambda)^{-1}\chi(A-\lambda)(A-\lambda)^{-1}u-(A-\lambda)^{-1}(A-\lambda)\chi(A-\lambda)^{-1}u]\\ &= \chi_1(A-\lambda)^{-1}[\chi,A](A-\lambda)^{-1}u\end{aligned}$$

and its counterparts for the realizations with boundary conditions. It is carefully expalined in [11] how this leads to the following theorem, where we use the terminology <u>spectrally negligible</u> for an operator family T_λ with the property: For any N and N' there is a constant $C_{N,N'}$ so that

$$(29)\qquad s_k(T_\lambda) \leq C_{N,N'}(1+|\lambda|)^{-N}k^{-N'} \quad \text{for } \lambda \text{ on a ray, } k=1,2,\ldots\ .$$

<u>Theorem 3</u>. *Let* $\chi \in C_0^\infty(\mathbb{R}^n)$, $\chi = 1$ *on a neighborhood of* Γ . *Then* $(1-\chi)G_\lambda$ *is spectrally negligible on each ray* $\lambda = re^{i\theta}$, $\theta \in]0,2\pi[$.

It should be noted that since we have taken 2m-order operators from the start, there is no need to prove the theorem specially for N-th powers of second order operators, as in [11].

The same technique allows us to replace G_λ by the analogous operator defined on a bounded set. More precisely, let Ω_1 be a smooth bounded open set containing $\bar{\Omega}_-$, and let A_1 resp. $A_{1,+}$ be the realizations of A on Ω_1 resp. on $\Omega_1 \cap \Omega_+$ determined by a new elliptic boundary condition at $\partial\Omega_1$ (the Dirichlet condition, for example) together with the <u>same boundary condition at</u> Γ <u>as</u> A_+ was defined from; we assume that A_1 and $A_{1,+}$ satisfy (1) on their domains. Define

$$G_{\lambda,1} = (A_1-\lambda)^{-1} - (A_{1,+}-\lambda)^{-1} \oplus (A_--\lambda)^{-1} , \tag{30}$$

where $\oplus$ now stands for the direct sum of operators defined on $L^2(\Omega_1 \cap \Omega_+)$ resp. $L^2(\Omega_-)$. We apply G_λ to functions in $L^2(\Omega_1)$ by extending them by zero on $\mathbb{R}^n \setminus \Omega_1$.

<u>Theorem 4</u>. *Let* $\chi \in C_0^\infty(\Omega_1)$, $\chi = 1$ *on a neighborhood of* $\bar{\Omega}_-$. *Then the operator* $\chi(G_\lambda - G_{\lambda,1})$ *on* $L^2(\Omega_1)$ *is spectrally negligible on each ray* $\lambda = re^{i\theta}$ *with* $\theta \in]0,2\pi[$.

The results allow us to restrict the attention to $G_{\lambda,1}$ defined entirely in terms of bounded domains. For a single domain with boundary one has the resolvent analysis of Seeley [21], but it is not quite sufficient for our purposes, since it does not include certain cross-terms in $G_{\lambda,1}$ arizing from extension by zero at one side of Γ, followed by application of $(A-\lambda)^{-1}$ and restriction to the <u>other</u> side of Γ. We here find it advantageous to view all the operators as special cases of the operators treated in the calculus of Boutet de Monvel [6], where the cross-terms are included as shown in Grubb [12]; then we determine the full structure of the entering operators in a rather easy way (this also gives a new proof of those results of Seeley that we use).

We study $G_{\lambda,1}$ in a local coordinate system, where the neighborhood of a piece of Γ is carried into a neighborhood of a piece of $\{x \in \mathbb{R}^n \mid x_n=0\}$, with Ω_+ going into $\{x_n > 0\}$. Let us denote by $I_\pm$ the function that equals 1 on $\mathbb{R}^n_\pm$ and 0 on $\mathbb{R}^n_\mp$, respectively. The operator $G_{\lambda,1}$ here takes the form (where we for simplicity write A, A_+ and A_- again for the transformed operators, and view the operators as extended to $\mathbb{R}^n$ resp. $\mathbb{R}^n_\pm$):

$$(31)\quad (A-\lambda)^{-1} - (A_+-\lambda)^{-1} \oplus (A_--\lambda)^{-1}$$

$$= (I_+ + I_-)(A-\lambda)^{-1}(I_+ + I_-) - (A_+-\lambda)^{-1} \oplus (A_--\lambda)^{-1}$$

$$= [I_+(A-\lambda)^{-1}I_+ - (A_+-\lambda)^{-1}] \oplus [I_-(A-\lambda)^{-1}I_- - (A_--\lambda)^{-1}]$$

$$+ I_+(A-\lambda)^{-1}I_- + I_-(A-\lambda)^{-1}I_+ .$$

The first term $I_+(A-\lambda)^{-1}I_+ - (A_+-\lambda)^{-1}$ (on $\mathbb{R}^n_+$) is the difference between the solution operator for the free space problem (the "fundamental solution"), cut down to $\mathbb{R}^n_+$, and the specific solution operator $(A_+-\lambda)^{-1}$ for the given boundary value problem (the "Green operator"). This difference, representing the adaptation to the particular boundary condition, is a simple example of a singular Green operator in the terminology of [6].

The second term $I_-(A-\lambda)^{-1}I_- - (A_--\lambda)^{-1}$ is a similar kind of operator on $\mathbb{R}^n_-$.

The last two terms in (31) are more strange, but here one finds (cf. Grubb [12]) that when they are composed with the reflection operator

$$J: u(x',x_n) \mapsto u(x',-x_n) ,$$

one gets operators

$$(32)\qquad I_+(A-\lambda)^{-1}I_-J \quad \text{and} \quad JI_-(A-\lambda)^{-1}I_+$$

that are singular Green operators on $\mathbb{R}^n_+$ in the sense of [6].

Now it is easy to find the structure of the symbols of these operators via the conventions in [6]. The symbol of A is a function $a(x,\xi)$ $(x \in \mathbb{R}^n, \xi \in \mathbb{R}^n)$ that is C^∞ in x and polynomial in ξ of degree $2m$. With $A-\lambda$ we associate the symbol $a(x,\xi) + \omega\mu^{2m}$, where $\mu = |\lambda|^{1/2m}$ and $|\omega| = 1$ $(\omega \neq -1)$; this is likewise a polynomial of order $2m$ (for each x and ω), now with respect to $(\xi,\mu) \subset \mathbb{R}^{n+1}$. The boundary conditions (which do not contain μ) are elliptic with respect to this symbol, when considered as constant in μ . Then $A-\lambda$, $A_+-\lambda$ and $A_--\lambda$ have parametrix symbols within the Boutet de Monvel calculus (in $n+1$ variables), satisfying the estimates and having the asymptotic expansions described in that theory. (One can find the symbols explicitly by classical calculations using the roots of the polynomial $a(x',0,\xi',\xi_n) + \omega\mu^{2m}$ in ξ_n , in local coordinates.) With this symbol description available, we can now give back to

μ the particular rôle of a parameter, and analyze the behavior of the symbols in their dependence on μ. The resulting estimates are sketched in the following theorem.

Theorem 5. *Consider* $G_{\lambda,1}$ *for* λ *on a ray disjoint from* $\mathbb{R}_+$ *(*$\lambda = re^{i\theta}$ *with* $\theta \in]0,2\pi[$*).*

1° $G_{\lambda,1}$ *is, in local coordinates near* Γ, *a sum of four terms of the following type (modulo reflection* $x_n \sim -x_n$*):*

$$(33)\quad Gu(x) = (2\pi)^{-n+1} \int_{\mathbb{R}^{n-1}} e^{ix'\cdot\xi'} \int_0^\infty \tilde{g}(x',x_n,y_n,\xi',\mu)\acute{u}(\xi',y_n)\,dy_n\,d\xi'$$

$(\acute{u}(\xi',x_n) = F_{x'\to\xi'}u(x',x_n))$, *where* $\tilde{g}(x',x_n,y_n,\xi',\mu)$ *is in* $\mathcal{S}(\overline{\mathbb{R}}_+\times\overline{\mathbb{R}}_+)$ *with respect to* (x_n,y_n), *satisfying the estimates for all indices:*

$$(34)\quad \|x_n^k D_{x_n}^{k'} y_n^m D_{y_n}^{m'} D_{x'}^{\beta} D_{\xi'}^{\alpha} D_\mu^j \tilde{g}\|_{L^2_{x_n,y_n}(\mathbb{R}^2_{++})}$$
$$\leq c(1+|\xi'|+|\mu|)^{-2m-k+k'-m+m'-|\alpha|-j}$$

for x' *in compact sets* $K \subset \mathbb{R}^{n-1}$; *here* c *depends on* K *and the indices. Moreover,* $\tilde{g}$ *is an asymptotic sum of quasi-homogeneous terms of decreasing order*

$$(35)\quad \begin{aligned} &\tilde{g} \sim \tilde{g}^0 + \tilde{g}^1 + \ldots + \tilde{g}^\ell + \ldots, \quad \text{where } \tilde{g}^\ell(x',x_n/t,y_n/t,t\xi',t\mu) = \\ &t^{-2m+1-\ell}\tilde{g}^\ell(x',x_n,y_n,\xi',\mu) \quad \text{for } |\xi'| \geq 1,\ t \geq 1, \end{aligned}$$

the lower order parts $\tilde{g} - \sum_{\ell<M} \tilde{g}^\ell$ *satisfying (34) with* $-2m$ *replaced by* $-2m-M$.

2° *If* $2m > n$ *(in particular, if* A *and* $A_\pm$ *are replaced by* A^N *and* $A_\pm^N$ *for some integer* N *with* $2mN > n$*), then* $G_{\lambda,1}$ *is of trace class, with an expansion*

$$(36)\qquad \operatorname{tr} G_{\lambda,1} \sim \sum_{j=1}^{\infty} d_j(-\lambda)^{-1+(n-j)/2m} \quad \text{for } \lambda = re^{i\theta},\ r\to\infty.$$

The structure of

$$B_{\lambda,1} = (A_1-\lambda)^{-1} - (A_{1,+}-\lambda)^{-1} \oplus 0$$

is likewise analyzed in this way (or one can combine the known results for $(A_--\lambda)^{-1}$ [2], [21], with the new analysis of $G_{\lambda,1}$), and we find in view of Theorems 3 and 4

Theorem 6. *Let* $2m > n$ *(or, if necessary, replace* A *and* $A_\pm$ *by* A^N *resp.* $(A_\pm)^N$ *with* $2mN > n$*). Then* G_λ *and* B_λ *are of trace class, the trace of* G_λ *has the* *same expansion* *as* $G_{\lambda,1}$ *in (36), and the trace of* B_λ *has an expansion*

$$(37) \qquad \operatorname{tr} B_\lambda \sim \sum_{j=0}^{\infty} c_j(-\lambda)^{-1+(n-j)/2m} \quad \text{for} \quad \lambda = re^{i\vartheta}, \quad r \to \infty,$$

where c_0 *is the same as the first coefficient in the similar expansion for* $(A_- - \lambda)^{-1}$.

One can moreover show that the trace norms of B_λ resp. G_λ are $O(|\lambda|^{-1+n/2m})$ resp. $O(|\lambda|^{-1+(n-1)/2m})$ for $\lambda \to -\infty$ on $\mathbb{R}_-$, cf. [11].

Now the deduction of Theorem 1 from Theorems 3-6 is practically covered by [11], so we need not go into details here. (Let us just add that the property $G_{\exp}(-t) \in \bigcap_{p>0} \mathfrak{S}_{(p)}$ for each $t > 0$ follows, as in [12], from the fact that $G_{\exp}(-t)$ is a singular Green operator of order $-\infty$; here $G_{\exp}(-t)$ is defined from G_λ by a Cauchy integral as described in [11], and the symbol is similarly defined from the symbol of G_λ.)

As for Theorem 2, only the trace class properties of $G^{(-s)}$ and $B^{(-s)}$ are shown in [11], so here we need to supply a proof. We use the Cauchy integral formula (for $\operatorname{Re} z < 0$), where C is a curve around $[r_0,\infty[$ (r_0 being a positive lower bound for the spectra of A, A_+ and A_-) and avoiding $\mathbb{R}_-$:

$$(38) \qquad G^{(z)} = \frac{i}{2\pi}\int_C \lambda^z G_\lambda \, d\lambda$$
$$= \frac{i}{2\pi}\int_C \lambda^z \chi G_{\lambda,1} d\lambda + \frac{i}{2\pi}\int_C \lambda^z S_\lambda d\lambda = G_1^{(z)} + G_2^{(z)};$$

where S_λ is spectrally negligible in view of Theorems 3 and 4. By use of the quasi-norm property (9), it is seen that the contribution from S_λ has rapidly decreasing characteristic values

$$(39) \qquad G_2^{(z)} \in \bigcap_{p>0} \mathfrak{S}_{(p)}.$$

For $G_1^{(z)}$ we observe that it can be pieced together from contributions described in local coordinates by formulas like (31), inserted in the Cauchy integral. Here

$$\frac{i}{2\pi}\int_C \lambda^z\chi[I_+(A_1-\lambda)^{-1}I_+ - (A_{1,+}-\lambda)^{-1}]d\lambda = \chi[I_+A^zI_+ - (A_{1,+})^z] ,$$

the "singular Green part" of $-\chi(A_{1,+})^z$, whereas

$$\frac{i}{2\pi}\int_C \lambda^z\chi I_+(A_1-\lambda)^{-1}I_-Jd\lambda \sim \chi I_+A^zI_-J ,$$

a "cross-term" arizing from A^z. Both terms are covered by the work of Laptev [15], the former because it is the singular Green part of a complex power of a differential boundary problem, the latter because it is a cross-term defined from a classical pseudo-differential operator A^z of order $\operatorname{Re} z < 0$ (cf. Seeley [20]). Altogether, we find by use of the analysis in [15]

$$s_k(G_1^{(z)}) \leq Ck^{-2m|\operatorname{Re} z|/(n-1)} \quad \text{for} \quad k = 1,2,\ldots . \tag{40}$$

The spectral estimates can also be investigated from the point of view of the analysis of singular Green operators in [12] and [13, Section 4.5] (the latter analysis is carried out in a very general pseudo-differential setting, where the best estimates are obtained only for $\operatorname{Re} z < -1/4m$). The rest of Theorem 2 follows from perturbation arguments given in [10] or [12].

Let us end by some remarks on other functions f . It is certainly possible to analyze G_f and B_f for functions $f(\lambda)$ such as λ^{is} $(s \in \mathbb{R})$ and $\exp(-t\lambda^s)$. On the other hand, the above points of view do not apply to functions like $\lambda^z\exp(it\lambda^{1/2m})$ without further techniques (and only for relatively small t , since the "propagation of singularities" interferes); a discussion is given at the end of [11].

BIBLIOGRAPHY

[1] S. Agmon: The coerciveness problem for integro-differential forms, J. Analyse Math. 6 (1958), 183-223.

[2] S. Agmon: Asymptotic formulas with remainder estimates for eigenvalues of elliptic operators. Arch. Rat. Mech. Anal. 28 (1968), 165-183.

[3] C. Bardos, J.C. Guillot and J. Ralston: La relation de Poisson pour l'équation des ondes dans un ouvert non borné. Application à la théorie de la diffusion, Comm. Part. Diff. Equ. 7 (1982), 905-958.

[4] M.S. Birman: Perturbations of the continuous spectrum of a singular elliptic operator under changes of the boundary and boundary conditions, Vestn. Leningr. 1 (1962), 22-55.

[5] M.S. Birman and M.Z. Solomiak: Estimates of singular numbers of integral operators, Russian Math. Surveys 32 (1977), 15-89.

[6] L. Boutet de Monvel: Boundary problems for pseudo-differential operators, Acta Math. 126 (1971), 11-51.

[7] P. Deift and B. Simon: On the decoupling of finite singularities from the question of asymptotic completeness in two body quantum systems, J. Functional Analysis 23 (1973), 218-238.

[8] P. Greiner: An asymptotic expansion for the heat equation, Arch. Rat. Mech. Anal. 41 (1971), 163-218.

[9] G. Grubb: On coerciveness and semiboundedness of general boundary problems, Israel J. Math. 10 (1971), 32-95.

[10] G. Grubb: Remainder estimates for eigenvalues and kernels of pseudo-differential elliptic systems, Math. Scand. 43 (1978), 275-307.

[11] G. Grubb: Remarks on trace estimates for exterior boundary problems, Comm. Part. Diff. Equ. 9 (1984), 231-270.

[12] G. Grubb: Singular Green operators and their spectral asymptotics, Duke Math. J. 51 (1984), 477-528.

[13] G. Grubb: Functional Calculus of Pseudo-differential Boundary Problems, monograph to appear in Birkhäuser Progress in Mathematics Series.

[14] A. Jensen and T. Kato: Asymptotic behavior of the scattering phase for exterior domains, Comm. Part. Diff. Equ. 3 (1978), 1165-1195.

[15] A.A. Laptev: Spectral asymptotics of a class of Fourier integral operators, Trudy Mosc. Math. Soc. 43 (1981) = Trans. Moscow Math. Soc. 1983, 101-127.

[16] J.L. Lions and E. Magenes: Problèmes aux limites non homogènes et applications, vol. 1. Editions Dunod, Paris 1968.

[17] A. Majda and J. Ralston: An analogue of Weyl's formula for unbounded domains III. An epilogue, Duke Math. J. 46 (1979), 725-731.

[18] H. McKean and I. Singer: Curvature and the eigenvalues of the Laplacian, J. Diff. Geometry 1 (1967), 43-69.

[19] M. Reed and B. Simon: Methods of Modern Mathematical Physics III, Academic Press, New York 1978.

[20] R. Seeley: Complex powers of an elliptic operator, A.M.S. Proc. Symp. Pure Math. 10 (1967), 288-307.

[21] R. Seeley: The resolvent of an elliptic boundary problem. Amer. J. Math. 91 (1969), 889-920.

[22] R. Seeley: Analytic extension of the trace associated with elliptic boundary problems, Amer. J. Math. 91 (1969), 963-982.

Commutator methods and asymptotic completeness for one - dimensional Stark effect Hamiltonians.

Arne Jensen*

Matematisk Institut
Aarhus Universitet
DK - 8000 Aarhus C
Denmark

and

Department of Mathematics
University of Kentucky
Lexington, KY 40506 - 0027
USA

Abstract.

Existence and strong completeness of the wave operators is shown for a new class of one - dimensional Stark effect Hamiltonians using a commutator method.

* Partially supported by NSF grant DMS - 8401748.

1. Introduction.

Let $H_0 = -\frac{d^2}{dx^2} + x$ denote the free Stark effect Hamiltonian in $L^2(\mathbb{R})$. H_0 is essentially selfadjoint on $\mathcal{S}(\mathbb{R})$, the Schwartz space of rapidly decreasing functions. Let $H = H_0 + V$, where V is multiplication by a real-valued function. We assume $V(x) = V_1(x) + V_2(x)$, where $V_1 = W''$, $W^{(j)} \in L^\infty(\mathbb{R})$, $j = 0,\ldots,4$, and $V_2 \in L^2_{loc}(\mathbb{R})$, with $V_2(x) = O(|x|^{-\delta})$ as $x \to -\infty$ for some $\delta > \frac{1}{2}$ and $V_2(x) = o(|x|)$ as $x \to +\infty$, in a suitable sense, see Assumption 3.1. This class of potentials is fairly large, in that the V_1 - part can be a periodic or almost - periodic function; see Section 4 for examples.

Our main result (Theorem 3.7) is existence of the wave operators

$$W_\pm = \underset{t \to \pm\infty}{s - \lim} e^{itH} e^{-itH_0}$$

and strong asymptotic completeness, i.e. the ranges of $W_\pm$ are equal to the subspace of absolute continuity of H, and the singular continuous spectrum is empty: $\sigma_{sc}(H) = \emptyset$. Furthermore, the point spectrum of H is discrete in $\mathbb{R}$.

The class of potentials V treated here is larger than those considered previously. Existence and completeness for $V = V_2$, $V_1 = 0$, has been shown previously by several authors, see e.g. [1,6,11,12,13, 15,16]. The case $V = V_1$, $V_2 = 0$, was treated in [8]. The absence of singular continuous spectrum has been shown for larger classes of potentials, see e.g. [2,3,12,15]. On the other hand, if V is too singular, it is known that the spectrum can be pure point. See [4] for recent results in case V is a randon Krönig - Penney potential.

The proof of asymptotic completeness is here based on a simple commutator computation and propagation estimates for e^{-itH_0} with respect to the conjugate operator $A = i\frac{d}{dx}$. These estimates are obtained by direct computation. The argument showing how to obtain asymptotic completeness from such estimates is due to Mourre [10], and is a version of the Enss Method (see [11] for extensive discussion of this method).

The main point of our proof of completeness is to show that V is short range with respect to A. The results given here generalize those in [8].

Let us briefly describe the organization of this paper. In Section 2 we prove propagation estimates for e^{-itH_0}, and also give some phase space propagation results. In Section 3 we show V is short range with respect to A, and then prove the main theorem. In Section 4 we give examples of potentials satisfying our assumptions.

2. Estimates for H_0.

Let $H_0 = -\frac{d^2}{dx^2} + x$ in $L^2(\mathbb{R})$. H_0 is essentially selfadjoint on $S(\mathbb{R})$. The operator $A = i\frac{d}{dx} = -p$ is essentially selfadjoint on $S(\mathbb{R})$, and is conjugate to H_0 at any point $E \in \mathbb{R}$, according to Mourre's definition [9], since we have

$$i[H_0, A] = I \tag{2.1}$$

on $S(\mathbb{R})$. The propagation estimates for e^{-itH_0} with respect to A can then be obtained from [9]. However, due to (2.1), there is an elementary direct proof. To state the result, let $P_A^+ (P_A^-)$ denote the spectral projection for A corresponding to $(0,\infty)$ $((-\infty,0))$.

<u>Proposition 2.1.</u> <u>For</u> $s \geq 0$ <u>the following estimates hold</u>:

$$\| (A^2+1)^{-s/2} e^{-itH_0} (A^2+1)^{-s/2} \| \leq c(1+|t|)^{-s}, \quad t \in \mathbb{R} \tag{2.2}$$

$$\| (A^2+1)^{-s/2} e^{-itH_0} P_A^{\pm} \| \leq c(1+|t|)^{-s}, \quad \pm t > 0 \tag{2.3}$$

<u>Proof</u>. The relation (2.1) implies on $S(\mathbb{R})$

$$e^{itH_0} A e^{-itH_0} = A + t ,$$

if we note that e^{-itH_0} maps $S(\mathbb{R})$ into itself. The estimates (2.2), (2.3) follow easily from this relation. □

The estimates (2.2), (2.3) lead to existence and completeness of the wave operators for H_0, $H_0 + V$, if V is short range with respect to A. Results of this type are implicit in [10]. An abstract scattering theory using a conjugate operator was given in [9]. We state the result in the following form needed below. The proof is similar to those in [10,11].

Proposition 2.2. Let V be multiplication by a realvalued function $V(x)$ such that $\mathcal{D}(V) \supseteq \mathcal{D}(H_0)$ and V is H_0 - bounded with bound less than one. Let $H = H_0 + V$. Assume

A.1) $(H+i)^{-1} - (H_0+i)^{-1}$ is compact.

A.2) There exists $\mu > 1$ such that the operator

$$(H+i)^{-2} V (H_0+i)^{-2} (A^2+1)^{\mu/2}$$

extends to a bounded operator on $L^2(\mathbb{R})$.

Then the wave operators

$$W_{\pm} = s - \lim_{t \to \pm\infty} e^{itH} e^{-itH_0}$$

exist and are complete, i.e. the ranges of $W_{\pm}$ are equal to the subspace of absolute continuity for H. Furthermore, the singular continuous spectrum of H is empty, and the point spectrum is discrete in $\mathbb{R}$.

We conclude this section with a short discussion of the phase space propagation properties of e^{-itH_0}. Let us first look at the corresponding classical problem (see [1]). We have chosen mass $m = \frac{1}{2}$ and strength of electric field $E = 1$ such that the Hamiltionian is $H(x,p) = p^2 + x$. The solutions to Hamilton's equations with $x(0) = x_0$, $p(0) = p_0$ are given by

$$x(t) = -t^2 + 2t\,p_0 + x_0$$

$$p(t) = -t + p_0 .$$

The corresponding properties for the quantum mechanical observables are given in the two propositions below. We use $F(p < p_0)$ to denote the spectral projection for the operator p corresponding to $(-\infty, p_0)$.

<u>Proposition 2.3.</u> <u>Let</u> $p_0 \in \mathbb{R}$ <u>be fixed. Then for all</u> $t \in \mathbb{R}$ <u>we have</u>

$$F(p > -t + p_0) e^{-itH_0} F(p < p_0) = 0$$

<u>and</u>

$$F(p < -t + p_0) e^{-itH_0} F(p > p_0) = 0 .$$

<u>Proof</u>. The results follow from the relation

$$e^{itH_0} p e^{-itH_0} = p - t$$

which is valid on $S(\mathbb{R})$.

□

As a special case we get with $A = -p$ the results:

$$P_A^- e^{-itH_0} P_A^+ = 0 \quad \text{for} \quad t > 0$$

$$P_A^+ e^{-itH_0} P_A^- = 0 \quad \text{for} \quad t < 0 .$$

These results show that $P_A^+ L^2(\mathbb{R})$ is the subspace of outgoing states, and $P_A^- L^2(\mathbb{R})$ is the subspace of incoming states, in the stricter sense of the Lax-Phillips scattering theory; cf. the discussion in [11].

The x - space properties of e^{-itH_0} follow from the lemma below which is somewhat sharper than the result obtained in [5; Lemma 2.7]. The notation $f(t) = 0(t^{-\infty})$ as $t \to \infty$ means that for any $N > 0$ there exists $c_N > 0$ such that

$$|f(t)| \leq c_N t^{-N}$$

for all $t > 0$.

<u>Lemma 2.4</u>. <u>Let</u> $\varphi \in C_0^\infty((-\infty, 0])$. <u>Then we have</u>

$$\| F(x > 0) e^{-itp^2} \varphi(p) F(x < 0) \| = 0(t^{-\infty})$$

<u>as</u> $t \to \infty$.

Proof. The result follows by a stationary phase argument, if we use that $p^{-k}(\frac{d}{dp})^j\varphi(p)$ is bounded for all $k \geq 0$, $j \geq 0$.

□

Proposition 2.5. Let $\varphi \in C_0^\infty((-\infty,o])$ and let $x_0 \in \mathbb{R}$, $p_0 \in \mathbb{R}$ be fixed. Then we have

$$\|F(x > -t^2 + 2tp_0 + x_0)e^{-itH_0}\varphi(p - p_0)F(x < x_0)\| = O(t^{-\infty}) \quad \text{as} \quad t \to +\infty$$

and

$$\|F(x > -t^2 + 2tp_0 + x_0)e^{-itH_0}\varphi(p_0 - p)F(x < x_0)\| = O(|t|^{-\infty}) \text{ as } t \to -\infty .$$

Proof. The idea is to reduce the result to the one in Lemma 2.4. Let

$$U = e^{-ip^3/3} \tag{2.4}$$

We have that U is unitary and

$$U H_0 U^{-1} = x$$

(multiplication operator), se [1]. We then get the representation

$$e^{-itH_0} = e^{-itx}e^{-itp^2}e^{it^2p}e^{-it^3/3}$$

which together with the translation properties

$$e^{-iax}g(p)e^{iax} = g(p+a)$$

$$e^{-iap}f(x)e^{iap} = f(x-a)$$

lead to the reduction. Details are omitted.

□

3. Asymptotic completeness for Stark effect Hamiltonians.

Let $B^k(\mathbb{R})$ denote the functions which are bounded with all derivations up to order k bounded. We prove asymptotic completeness for the following class of potentials. χ_I denotes the characteristic function for the interval I.

Assumption 3.1. *Let* V *be a realvalued function such that* $V(x) = V_1(x) + V_2(x)$, *where these functions satisfy:*

V.1) *There exists* $W \in B^4(\mathbb{R})$ *such that* $V_1 = W''$,

V.2) $\mathcal{D}(V_2) \supseteq \mathcal{D}(H_0)$ *and* $V_2(H_0 + i)^{-1}$ *is compact. There exists* $\delta > 1/2$ *such that* $(1 - x^2)^{\delta/2} \chi_{(-\infty,0)}(x) V_2(H_0 + i)^{-1}$ *is bounded.*

Examples of potentials satisfying Assumption 3.1 will be given in Section 4.

We need the following result, due to Herbst - Simon.

Lemma 3.2. *The following operators extend to bounded operators on* $L^2(\mathbb{R})$.

$$x \chi_{[0,\infty)}(x)(H_0 + i)^{-1} , \tag{3.1}$$

$$\sqrt{x}\, \chi_{[0,\infty)}(x)\, p\, (H_0 + i)^{-1} , \tag{3.2}$$

$$\chi_{[0,\infty)}(x) p^2 (H_0 + i)^{-1} . \tag{3.3}$$

Proof. Se [7; Theorem c.1]. □

In the following sequence of lemmas it is shown that potentials satisfying Assumption 3.1 are short range with respect to A , i.e. satisfy A.2) in Proposition 2.2.

Lemma 3.3. Let U be multiplication by a function $U(x)$ with $\mathcal{D}(U) \supseteq \mathcal{D}(H_0)$. Assume that for some δ with $\frac{1}{2} < \delta \leq 1$ the operator

$$(1+x^2)^{\delta/2}\, U(H_0+i)^{-1}$$

is bounded. Then the operator

$$U(H_0+i)^{-2}(A^2+1)^{\delta}$$

extends to a bounded operator on $L^2(\mathbb{R})$.

Proof. Let W be a multiplication operator with $\mathcal{D}(W) \supseteq \mathcal{D}(H_0)$ and $W(H_0+i)^{-1}$ bounded. We have

$$(1+x^2)^{-\frac{1}{2}}\, W p^2 (H_0+i)^{-2}$$

$$= (1+x^2)^{-\frac{1}{2}}\, W H_0 (H_0+i)^{-2} - (1+x^2)^{-\frac{1}{2}}\, W x (H_0+i)^{-2}\,.$$

From this expression we get

$$\| (1+x^2)^{-\frac{1}{2}}\, W(p^2+1)(H_0+i)^{-2} \| \leq c\, \| W(H_0+i)^{-1} \|\,.$$

Using $[H_0, p^2] = -2ip$ one gets after a straightforward computation (see also the proof of Lemma 3.4)

$$\| (1+x^2)^{-\frac{1}{2}}\, W(H_0+i)^{-2}(p^2+1) \| \leq c\, \| W(H_0+i)^{-1} \|\,.$$

An interpolation argument gives for δ, $0 \leq \delta \leq 1$:

$$\| (1+x^2)^{-\delta/2}\, W(H_0+i)^{-2}(p^2+1)^{\delta} \| \leq c\, \| W(H_0+i)^{-1} \|\,.$$

Taking $W = (1+x^2)^{\delta/2}\, U$ the result in the lemma follows. □

Lemma 3.4. Let U be multiplication by a function $U(x)$ such that $U(x) = 0$ for $x \leq 0$ and $\mathcal{D}(U) \supseteq \mathcal{D}(H_0)$, U H_0-bounded with bound less than one. Then the operator

$$(H_0+i)^{-1}\, U(H_0+i)^{-1}(A^2+1)$$

extends to a bounded operator on $L^2(\mathbb{R})$.

Proof. We have $\chi_{[0,\infty)}(x)U(x) = U(x)$ and thus

$$(H_0 + i)^{-1} U p^2 (H_0 + i)^{-1}$$

$$= (H_0 + i)^{-1} U \chi_{[0,\infty)}(x) p^2 (H_0 + i)^{-1} .$$

The result (3.3) shows that this operator extends to a bounded operator on $L^2(\mathbb{R})$. On $S(\mathbb{R})$ we have

$$(H_0 + i)^{-1} U (H_0 + i)^{-1} (p^2 + 1)$$

$$= (h_0 + i)^{-1} U (p^2 + 1)(H_0 + i)^{-1}$$

$$- 2i (H_0 + i)^{-1} U (H_0 + i)^{-1} p (H_0 + i)^{-1} .$$

We note that $(H_0 + i)^{-1}$, $(p + i)^{-1}$ etc. all map $S(\mathbb{R})$ into $S(\mathbb{R})$. Thus we get

$$(H_0 + i)^{-1} U (H_0 + i)^{-1} (p^2 + 1)\{1 + 2i(p + 4i)^{-1} (H_0 + i)^{-1}\}$$

$$= (H_0 + i)^{-1} U (p^2 + 1)(H_0 + i)^{-1}$$

$$+ 2i (H_0 + i)^{-1} U (H_0 + i)^{-1} (p + 4i)^{-1} (H_0 + i)^{-1}$$

$$+ 8 (H_0 + i)^{-1} U (H_0 + i)^{-1} p (p + 4i)^{-1} (H_0 + i)^{-1} .$$

Since the operator in $\{\ldots\}$ is invertible and all terms on the right hand side extend to bounded operators on $L^2(\mathbb{R})$ the result in the Lemma follows. □

Lemma 3.5. Let V satisfy Assumption 3.1. Let $H = H_0 + V$. Then the operators

$$(H + i)^{-1} V_1 (H_0 + i)^{-1} p$$

and

$$(H + i)^{-2} V_1 (H_0 + i)^{-2} p^2$$

extend to bounded operators on $L^2(\mathbb{R})$.

Proof. The proof is a modified version of the one given in [8]. The extra complication comes from the term V_2 in the potential, since V_2 is not assumed differentiable. As in the proofs of Lemmas

3.3 and 3.4 it suffices to show that $(H+i)^{-1} V_1 \, p(H_0+i)^{-1}$ and $(H+i)^{-2} V_1 \, p^2(H_0+i)^{-2}$ extend to bounded operators on $L^2(\mathbb{R})$.

Let $U \in B^2(\mathbb{R})$. As a quadratic form on $S(\mathbb{R}) \times S(\mathbb{R})$ we find

$$i\,[H_0, U] = 2U'p - i\,U'' .$$

Again as a quadratic form on $S(\mathbb{R}) \times S(\mathbb{R})$ we compute

$$\begin{aligned} \frac{d}{ds} e^{isH} U e^{-isH_0} &= e^{isH}(i\,H\,U - i\,U\,H_0)e^{-isH_0} \\ &= e^{isH}(i\,V\,U + 2\,U'\,p - i\,U'')\,e^{-isH_0} . \end{aligned} \tag{3.4}$$

We take $U = W'$ from Assumption 3.1 V.1) and find $U' = V_1$. From (3.4) we get (with $s = 0$)

$$\begin{aligned} &(H+i)^{-1} V_1 \, p(H_0+i)^{-1} \\ &= (H+i)^{-i}\{i\,H\,U - iU\,H_0 - i\,V\,U + i\,U''\}(H_0+i)^{-1} . \end{aligned}$$

By Assumption 3.1 $V(H_0+i)^{-1}$ is bounded, hence the right hand side extends to a bounded operator on $L^2(\mathbb{R})$ This proves the first part of the lemma.

We now choose $U = W$ and differentiate once more in (3.4) to get

$$\frac{d^2}{ds^2} e^{isH} W e^{-isH_0} = \frac{d}{ds} e^{isH}\{i\,V\,W + 2W'p - i\,W''\}\,e^{-isH_0} \tag{3.5}$$

We multiply by $(H+i)^{-2}$ from the left and by $(H_0+i)^{-2}$ from the right in the above equation and discuss the terms separately. On the left hand side we clearly get a bounded operator on $L^2(\mathbb{R})$. The first term on the right hand side of (3.5) is given by

$$\frac{d}{ds} e^{isH}(H+i)^{-2}\; V\,W(H_0+i)^{-2}\,e^{-isH_0}$$

which clearly extends to a bounded operator on $L^2(\mathbb{R})$.

The second term in (3.5) is rewritten:

$$\begin{aligned} \frac{d}{ds} e^{isH}\, 2\,W'\,p\,e^{-isH_0} &= e^{isH}(i\,H\,2W'\,p - i\,2W'\,p\,H_0)e^{-isH_0} \\ &= e^{isH}(2\,i\,V\,W'\,p - 2W' + 4W''\,p^2 - 2\,i\,W^{(3)}p)e^{isH_0} \end{aligned} \tag{3.6}$$

We multiply by $(H+i)^{-2}$ from the left and $(H_0+i)^{-2}$ from the right in (3.6). The term $VW'p$ is written as

$$V_1 W'p + V_2 W'p = W''W'p + V_2 W'p$$
$$= \frac{1}{2}((W')^2)'p + V_2 W'p .$$

The first term is of the form U' considered in the first part of the proof, hence gives a bounded term. Since W' is a bounded function $V_2 W'$ will satisfy the conditions in Assumption 3.1 V.2), and we can use Lemmas 3.3, 3.4 to conclude that

$$(H+i)^{-2} V_2 W'p(H_0+i)^{-2}$$

extends to a bounded operator on $L^2(\mathbb{R})$. The term W' in (3.6) clearly gives a bounded operator. Since $W \in B^4(\mathbb{R})$, we can use the first part of the proof to conclude that $W^{(3)}p$ gives a bounded operator.

The third term in (3.4) is rewritten:

$$\frac{d}{ds} e^{isH}(-iW'')e^{-isH_0} = e^{isH}(VW'' - 2iW^{(3)}p + W^{(4)})e^{-isH_0} .$$

As above we easily see that multiplied by $(H+i)^{-2}$ and $(H_0+i)^{-2}$ from the left and right, respectively, we get a bounded operator.

Thus we can isolate the term $W''p^2 = V_1 p^2$ in (3.6) and repeat the argument in the first part of the proof. This proves the second part of the lemma. □

Lemma 3.6. *Let* V *satisfy Assumption 3.1 and let* $H = H_0 + V$. *Then*

$$(H+i)^{-1} - (H_0+i)^{-1}$$

is compact.

Proof. We have

$$(H+i)^{-1} - (H_0+i)^{-1} = (H+i)^{-1} V_1 (H_0+i)^{-1} + (H+i)^{-1} V_2 (H_0+i)^{-1} .$$

The second term is compact by Assumption 3.1 V.2). Lemma 3.5 and complex interpolation imply that

$$(H + i)^{-1} V_1 (H_0^2 + 1)^{-s/2} (p^2 + 1)^{s/2}$$

is bounded for $0 \leq s \leq 1$. We write

$$(H + i)^{-1} V_1 (H_0 + i)^{-1} = ((H + i)^{-1} V_1 (H_0^2 + 1)^{-1/4} (p^2 + 1)^{1/4})$$
$$((p^2 + 1)^{-1/4} (H_0^2 + 1)^{-1/4}) \; ((H_0^2 + 1)^{1/2} (H_0 + i)^{-1})$$

The first and last term are bounded. The middle term is unitarily equivalent via U defined in (2.4) to $(p^2 + 1)^{-1/4} (x^2 + 1)^{-1/4}$, hence compact, see e.g. [14]. □

With these preparations we have our main result:

Theorem 3.7. *Let* V *satisfy Assumption 3.1. Then the wave operators*

$$W_{\pm} = s - \lim_{t \to \pm \infty} e^{itH} e^{-itH_0}$$

exist and are asymptotically complete. Furthermore, $\sigma_{sc}(H) = \emptyset$ *and the point spectrum of* H *is discrete in* $\mathbb{R}$.

Proof. We use Proposition 2.2. It is clear from Assumption 3.1 that $\mathcal{D}(V) \supseteq \mathcal{D}(H_0)$ and that V is H_0 - bounded with bound equal to zero. Thus $H = H_0 + V$ is selfadjoint with $\mathcal{D}(H) = \mathcal{D}(H_0)$. Lemma 3.6 shows that condition A.1) in Proposition 2.2 is verified. Combining the results in Lemmas 3.3, 3.4, 3.5 we see that condition A.2) in Proposition 2.2 is verified with $\mu = 2\delta$, $\delta > \frac{1}{2}$ from Assumption 3.1. The result now follow from Proposition 2.2. □

4. Examples of potentials.

In this section we briefly give some examples of potentials satisfying Assumption 3.1. We first consider the V_1 part. We assume that $V_1(x)$ is a realvalued function which can be represented as

$$V_1(x) = \int_{-\infty}^{\infty} e^{i\omega x}\, d\mu(\omega) \quad ,$$

where μ is a complex Borel measure which satisfies

$$\int_{-\infty}^{\infty} (\omega^2 + \omega^{-2})\, d|\mu|(\omega) < \infty \ .$$

V_1 satisfies the assumption V.1). It suffices to take

$$W(x) = \int_{-\infty}^{\infty} (-\omega^{-2})\, d\mu(\omega) \quad .$$

We get a large class of almost - periodic functions, if we take μ to be a sum of point measures.

Another class of potentials satisfying V.1) is obtained, if we take

$$V_1 = \sum_{j=1}^{N} V_1^j$$

where each V_1^j is a periodic realvalued function with average zero over a period, and $V_1^j \in C^3(\mathbb{R})$. This can easily be seen by expanding each V_1^j in a Fourier series.

To give an example of the V_2 - part we take a realvalued function $V_2(x)$ which can be written as

$$V_2(x) = ((1+x^2)^{-\delta/2} \chi_{(-\infty,0]}(x) + (1+x^2)^{\frac{1}{2}} \chi_{[0,\infty)}(x))\,(U_1(x) + U_2(x))$$

where $\delta > \frac{1}{2}$, and where

$$U_1 \in L^\infty(\mathbb{R}) \quad \text{with} \quad \lim_{x\to\infty} U_1(x) = 0$$

and

$$U_2 \in L^2_{loc}(\mathbb{R}) \quad \text{with} \quad \lim_{|x|\to\infty} (1+x^2) \int_{|x-y|\le 1} |V(y)|^2\, dy = 0 \ .$$

It is straightforward to verify that V_2 satisfies condition V.2). See e.g. [16].

References.

1. Avron, J. E., Herbst, I. W.: Spectral and scattering theory for Schrödinger operators related to the Stark effect. Comm. Math. Phys. 52 (1977), 239 - 254.

2. Ben-Artzi, M. : Remarks on Schrödinger operators with an electric field and deterministic potentials. J. Math. Anal. Appl. 109 (1985), 333 - 339.

3. Bentosela, F., Carmona, R., Duclos, P., Simon, B., Souillard, B., Weder, R.: Schrödinger operators with an electric field and random or deterministic potentials. Comm. Math. Phys. 88 (1983), 387 - 397.

4. Delyon, F., Simon, B., Souillard, B.: From power pure point to continuous spectrum in disordered systems. Ann. Inst. H.Poincaré, Sect.A., 42 (1985), 283 - 309.

5. Enss, V.: Propagation properties of quantum scattering states. J. Funct. Anal. 52 (1983), 219 - 251.

6. Herbst, I. W.: Unitary equivalence of Stark effect Hamiltonians. Math. Z. 155 (1977), 55 - 70.

7. Herbst, I. W., Simon, B.: Dilation analyticity in constant electric fields. II. N - body problem, Borel summability. Comm. Math. Phys. 80 (1981), 181 - 216.

8. Jensen, A.: Asymptotic completeness for a new class of Stark effect Hamiltonians. Comm. Math. Phys., to appear.

9. Jensen, A., Mourre, E., Perry, P.: Multiple commutator estimates and resolvent smoothness in quantum scattering theory. Ann. Inst. H.Poincaré, Sect.A., 41 (1984), 207 - 224.

10. Mourre, E.: Link between the geometrical and the spectral transformation approches in scattering theory. Commun. Math. Phys. 68 (1979), 91 - 94.

11. Perry, P. A.: Scattering theory by the Enss method.
Math. Reports, 1, part 1, 1983.

12. Rejto, P. A., Sinha, K.: Absolute continuity for a 1-dimensional model of the Stark-Hamiltonian.
Helv. Phys. Acta. 49 (1976), 389-413.

13. Simon, B.: Phase space analysis of simple scattering systems: Extensions of some work of Enss.
Duke Math. J. 46 (1979), 119-168.

14. Simon, B.: Trace ideals and their applications.
Cambridge University Press, Cambridge 1977.

15. Veselic, K., Weidmann, J.: Potential scattering in a homogeneous electrostatic field.
Math. Z. 156 (1977), 93-104.

16. Yajima, K.: Spectral and scattering theory for Schrödinger operators with Stark-effect.
J. Fac. Sci. Univ. Tokyo, Sect. I A, 26 (1979), 377-390.

Lorentz Invariant Quantum Theory

Lars-Erik Lundberg
Matematisk Institut, HCØ
Universitetsparken 5
DK 2100 Copenhagen
Denmark

1. Introduction.

We shall consider a new type of quantum theory where the symmetry is governed by the homogeneous Lorentz group. In particular we shall consider a Lorentz invariant generalization of many-body Schrödinger theory, which has "electromagnetic" and "gravitational" interactions.

There are only fermion degrees of freedom in the theory, i.e. "electromagnetism" and "gravitation" are due to direct interaction among the fermions.

It is well-known that Maxwell's theory for electromagnetism can be obtained from electrostatics by imposing appropriate transformation properties under Lorentz transformations.

We shall follow an old suggestion by Pauli [1] and express the Schrödinger Hamiltonian with electrostatic interactions in the following form (in modern notation)

$$H = H_0 + \frac{1}{2}\int_{\mathbb{R}^3} :\nabla V(x)\nabla(x): \, dx \ ,$$

where $H_0 = \frac{1}{2m}\int_{\mathbb{R}^3} \nabla\phi^*(x)\nabla\phi(x)dx$ is the free Hamiltonian, $V(x) = \frac{e}{4\pi}\int \frac{\phi^*(y)\phi(y)}{|x-y|} dy$ is the Coulomb field and $\phi(x)$ is the charged fermion field. The analogy with electrostatics is striking when the Hamiltonian is written in this form and there are Lorentz covariant generalizations analogous to the transition from electrostatics to Maxwell's theory.

An essential ingredient in the construction, is the use of some results on wave equations with light-cone data.

In this lecture we shall concentrate on the (partly formal) construction of the Lorentz invariant quantum theory, and only superficially discuss the vast number of mathematical and physical questions that naturally arise.

In order to familiarize ourself with the new conceptual framework, we start out with a reformulation of Schrödinger theory the electrostatic interactions, which includes the Pauli form of the Hamiltonian, and gives rigorous meaning to the formulas above.

2. Galilei invariant Schrödinger theory.

We shall consider Schrödinger theory from a different angle than is usually done. In particular we shall focus on transformation properties under the homogeneous Galilei group. This is essential for the physical interpretation of the Lorentz invariant theory, and the translational invariance will be considered as an accidental symmetry of the non-relativistic theory.

2.1. State space, Galilei motions and position operators.

Let us consider spinless fermions, i.e. the one-body Hilbert space is given by $H_1 = L_2(\mathbb{R}^3)$, where we identify $\mathbb{R}^3$ with the one-body "momentum-space".

Let $\Lambda^n H_1$ denote the n-fold exterior power of H_1, i.e. $\Lambda^n H_1 \subset \otimes^n H_1 = L_2(\mathbb{R}^3 \times \cdots \times \mathbb{R}^3)$, and $\Lambda^n H_1$ consists of anti-symmetric square integrable functions in the variables $(k_1,\ldots,k_n)$, $k_i \in \mathbb{R}^3$, $i=1,\ldots,n$. The exterior product of n vectors $f_i \in H_1$, $i=1,\ldots,n$, is defined by $f_1 \Lambda \cdots \Lambda f_n = \frac{1}{\sqrt{n!}} \sum_{\sigma \in S_n} \chi(\sigma)\, f_{\sigma(1)} \otimes \cdots \otimes f_{\sigma(n)}$, where $\chi(\sigma)$ is the sign of the permutation σ of the symmetric group S_n. This means that if $\{e_i\}_{i \in \mathbb{N}}$ is an orthonormal basis for H_1 then $\{e_{i_1} \Lambda \cdots \Lambda e_{i_n}\}$, $i_1 < i_2 < \cdots < i_n$ gives an orthonormal basis for $\Lambda^n H_1$.

The anti-symmetric Fock-space H which we also will call the state space is given by

$$H = \bigoplus_{n=0}^{\infty} \Lambda^n H_1 , \quad \Lambda^0 H_1 = \mathbb{C} ,$$

and has a distinguished vector Ω, the Fock vacuum, given by $\Omega = \bigoplus_{n=0}^{\infty} \Omega_n$, $\Omega_0 = 1$, $\Omega_n = 0$, $n \geq 1$.

There is a natural complex linear map $f \to a(f)^*$ of H_1 into operators on H, such that the restriction of $a(f)^*$ to product vectors is given by

$$a(f)^*\Omega = f , \quad a(f)^*(f_1 \Lambda \cdots \Lambda f_n) = f \Lambda f_1 \Lambda \cdots \Lambda f_n ,$$

with the natural identification of vectors in $\Lambda^n H_1$ with vectors in H. In fact it follows from this that $\frac{1}{\|f\|} a(f)^*$ is a partial isometry and that $a(f)^*$ together with its adjoint fulfill the canonical anti-commutation relations (CAR)

$$a(f)a(g)^* + a(g)^*a(f) = \langle f,g \rangle 1 \quad , \quad a(f)a(g) + a(g)a(f) = 0 ,$$

for all $f,g \in H_1$ and $a(f)$ annihilate the Fock-vacuum, i.e. $a(f)\Omega = 0$ for all $f \in H_1$. This explicit representation of the CAR is called the Fock representation.

Let $\mathcal{D} \subset \mathcal{H}$ consist of those vectors $F = \bigoplus_{n=0}^{\infty} F_n$, such that only finitely many $F_n \neq 0$ and $F_n \in \mathcal{S}(\mathbb{R}^{3n})$ (the Schwartz space). Let us for $k \in \mathbb{R}^3$ define $\alpha(k): \mathcal{D} \to \mathcal{D}$ by the following formula

$$(\alpha(k)F_n)(k_1,\ldots,k_{n-1}) = \sqrt{n}\, F_n(k,k_1,\ldots,k_{n-1}) .$$

This densely defined operator $\alpha(k)$ is related to $a(f)$, in fact

$$a(f) = \int_{\mathbb{R}^3} \overline{f(k)}\alpha(k)\,dk ,$$

on $\mathcal{D}$. The formal adjoint $\alpha(k)^*$ does only define a sesquilinear form on $\mathcal{D} \times \mathcal{D}$ by

$$< F,\alpha(k)^* G > \equiv < \alpha(k)F,G > ,$$

for $F,G \in \mathcal{D}$. In fact when properly interpreted one gets $\alpha(k)\alpha(k')^* + \alpha(k')^*\alpha(k) = \delta(k-k')1$, where the δ is a 3-dimensional Dirac delta function.

The operator $\alpha(k): \mathcal{D} \to \mathcal{D}$ can be Fourier transformed i.e.

$$\phi(x) = (2\pi)^{-3/2} \int_{\mathbb{R}^3} e^{ik\cdot x}\,\alpha(k)\,dk ,$$

defines an operator $\phi(x): \mathcal{D} \to \mathcal{D}$ and this operator will be called the <u>Schrödinger field</u>.

Many of the operators occuring in quantum theory can easily be expressed in terms of the $\alpha(k)$'s or the $\phi(x)$'s, and this motivates the introduction of these singular objects. The free Schrödinger Hamiltonian H_0 can be expressed as

$$H_0 = \int \frac{|k|^2}{2m}\,\alpha(k)^*\alpha(k)\,dk = \frac{1}{2m}\int \nabla\phi(x)^*\nabla\phi(x)\,dx ,$$

when considered as a sesquilinear form. The number operator N can be written as

$$N = \int \alpha(k)^*\alpha(k)\,dk = \int \phi(x)^*\phi(x)\,dx ,$$

and the momentum operators $P = (P_1,P_2,P_3)$ as

$$P = \int k\alpha(k)^*\alpha(k)\,dk = -i\int \phi(x)^*\nabla\phi(x)\,dx .$$

We shall now consider the action of the homogeneous Galilei group G_0 on the Fock-space $\mathcal{H}$. The action of G_0 on the "momentum-space" $\mathbb{R}^3$ is that given by the Euclidean group of motions, i.e. G_0's action on $\mathbb{R}^3$ is generated by translations $k \to k + mv$, $m > 0$, $v \in \mathbb{R}^3$ (which also will be called Galilei boosts) and rotations $k \to rk$, $r \in SO(3)$.

There is a natural unitary representation $g \to U_1(g)$ of G_0 on $\mathcal{H}_1 = L_2(\mathbb{R}^3)$ given by

$$(U_1(g)f)(k) = f(g^{-1}k) ,$$

and this representation lifts to a unitary representation $g \to U(g)$ of G_0 on the Fock-space $\mathcal{H}$. It is enough to give the action of $U(g)$ on product vectors in $\mathcal{H}$

$$U(g)\Omega = \Omega , \quad U(g)(f_1 \wedge \cdots \wedge f_n) = U_1 f_1 \wedge \cdots \wedge U_1 f_n .$$

The free Hamiltonian H_0 is not invariant under the unitary motion group $g \to U(g)$. However if we introduce the <u>free Schrödinger mass operator</u> M_0 by

$$M_0 = mN + H_0 - \frac{P^2}{2mN} ,$$

then one can show that M_0 is invariant, i.e.

$$U(g)M_0U(g)^{-1} = M_0 \quad \text{for all} \quad g \in G_0 . \qquad (*)$$

The mass operator M_0 is the generator of the free internal or relative dynamics, where we have included the "free mass term" mN for later convenience.

Let $g_v \in G_0$ be defined by $g_v : k \to k + mv$. The generators $B = (B_1, B_2, B_3)$ of $U(g_v)$ are given by

$$B = -im \int \alpha(k)^* \nabla\alpha(k)\, dk = m \int x\phi(x)^* \phi(x)\, dx$$

and they commute with M_0 due to (*). The generators B are up to a mass, the center of mass position operators, and these are invariant under the free internal or relative dynamics.

Let $F_n \in \Lambda^n \mathcal{H}_1$ and consider the map

$$F_n(k_1, \ldots, k_n) \to F_n(g^{-1}k_1, k_2, \ldots, k_n) ,$$

i.e. $g \in G$ only acts one variable. We denote this unitary map by $U_{1n}(g)$ and the generators for $U_{1n}(g_v)$ are up to the constant m just the position operator associated with the variable k_1. Analogously we can introduce relative position operators. All this means that the notion of position operators in Schrödinger theory can be introduced, also when one only considers homogeneous Galilei transformations.

If we add space translations to G_0 we get a group which is not a unitary symmetry of Schrödinger theory and we feel that the space-translational invariance of Schrödinger theory might be an accidential symmetry. The unitary symmetry under G_0 is considered to be the fundamental symmetry and reflects the Lorentz group symmetry of a "relativistic" more fundamental theory.

2.2. The electrostatic Hamiltonian H_S and mass operator M_S.

Let us now introduce a repulsive electrostatic interaction among the particles. The standard expression for the electrostatic Hamiltonian H_S is given by

$$H_S = H_0 + \frac{1}{2} \int \phi(x)^* \phi(x')^* \frac{e^2}{|x-x'|} \phi(x')\phi(x)\,dx dx' ,$$

where e is the electric charge of the electron. This expression defines a non-negative symmetric sesquilinear form on $\mathcal{D} \times \mathcal{D}$ and therefore a self-adjoint operator (due to Friedrich's theorem). We shall rewrite this expression as indicated in the introduction.

The Coulomb field $V(x)$ is defined by the following non-negative sesquilinear form on $\mathcal{D} \times \mathcal{D}$

$$V(x) = \frac{e}{\sqrt{4\pi}} \int \frac{\phi(y)^* \phi(y)}{|x-y|}\,dy ,$$

i.e. $V(x)$ defines a self-adjoint operator in $\mathcal{H}$.

In 1933 Pauli [1] noted that the electrostatic Hamiltonian H_S can be written on the following form

$$H_S = H_0 + \frac{1}{2} \int :\nabla V(x) \nabla V(x):\,dx ,$$

where the double dots $:\ :$ stands for normal ordering, i.e. all "operators ϕ^*" are anti-commuted to the left. Pauli did not have the appropriate ordering, and this caused infinite self-anergy. Pauli suggests however that this form of H_S could be the starting point for generalizations including retardation.

Let us express the interaction Hamiltonian in terms of the α's. A simple computation gives

$$\frac{1}{2} \int :\nabla V \nabla V:\,dx = \frac{e^2}{2} \int \alpha(k_1)^* \alpha(k_2)^* \frac{4\pi\delta(k_1+k_2-k_3-k_4)}{(2\pi)^3 |k_1-k_3|^2} \alpha(k_3)\alpha(k_4)\,dk_1 \cdots dk_4 ,$$

where δ is the 3-dimensional Dirac "delta function".

This interaction Hamiltonian is invariant under the unitary motion group $g \to U(g)$. The invariant electrostatic mass operator M_S is given by

$$M_S = mN + H_S - \frac{P^2}{2mN} ,$$

and it generates the internal or relative dynamics, which is non-trivial part of the dynamics.

3. Lorentz invariant quantum theory.

We shall consider Lorentz covariant generalizations of the electrostatic Schrödinger operator H_S and mass operator M_S. The approach is analogous to the transition from classical electrostatics to Maxwell's theory, but the theories obtained are not Poincaré covariant, contrary to the classical case. We suggest that this breaking of Poincaré invariance might be essential in "relativistic" quantum theories.

3.1. State space, motions and position operators.

Let us start with the construction of the free Dirac field and this will be considered as a relativistic generalization of the Schrödinger field.

The free Dirac field will transform properly under the full Poincaré group, and it is only later when we introduce the interaction, the symmetry is reduced to the Lorentz group.

The construction of the "one-particle" Hilbert-space H_1 is closely related to the Dirac wave equation and also to spin 1/2 irreducible representations of the Poincaré group.

Let L denote the proper Lorentz group, i.e. linear maps $\Lambda : \mathbb{R}^4 \to \mathbb{R}^4$, with $\det \Lambda = 1$, which leaves the quadratic form $x_0^2 - x_1^2 - x_2^2 - x_3^2$ and the positive cone $x_0^2 - x_1^2 - x_2^2 - x_3^2 = 0, x_0 > 0$ invariant.

We shall in the following consider $SL(2,\mathbb{C})$, the group of complex 2×2 matrices with determinant one. This is the simply connected covering of L and will for simplicity be denoted by G.

Let us give the connection between the proper Lorentz group L and $G = SL(2,\mathbb{C})$.

For $x = (x_0, x_1, x_2, x_3) \in \mathbb{R}^4$ we define

$$a(x) = \begin{pmatrix} x_0+x_3 & x_1-ix_2 \\ x_1+ix_2 & x_0-x_3 \end{pmatrix} = x_0 1 + \sum_{i=1}^{3} x_i\sigma_i ,$$

where σ_i are the so-called Pauli matrices

$$\sigma_1 = \begin{pmatrix} 0 & 1 \\ 1 & 0 \end{pmatrix}, \quad \sigma_2 = \begin{pmatrix} 0 & -i \\ i & 0 \end{pmatrix}, \quad \sigma_3 = \begin{pmatrix} 1 & 0 \\ 0 & -1 \end{pmatrix}.$$

The 2×2 matrix $a(x)$ is symmetric and every symmetric 2×2 matrix is of the form $a(x)$ for some $x \in \mathbb{R}^4$. This means that the map $a(x) \to g\,a(x)g^*$, $g \in G$ defines a linear map $\Lambda_g : \mathbb{R}^4 \to \mathbb{R}^4$ and it follows easily that $\Lambda_g \in L$. We shall in the following refer to G as the Lorentz group.

The Dirac wave equation is a system of coupled linear partial differential equations on $\mathbb{R}^4$, which can be written

$$\left(i \sum_{\mu=0}^{3} \gamma_\mu \frac{\partial}{\partial x_\mu} - m \right) u(x) = 0 ,$$

with $u : \mathbb{R}^4 \to \mathbb{C}^4$, $m > 0$ and γ_μ, $\mu = 0,\dots 3$ are four complex 4×4 matrices, called Dirac gamma matrices fulfilling $\gamma_\mu\gamma_\nu + \gamma_\nu\gamma_\mu =$

$$\begin{cases} +I, & \mu = \nu = 0 \\ -I, & \mu = \nu \geq 1 \\ 0, & \mu \neq \nu \end{cases}.$$

An explicit representation is given by

$$\gamma_0 = \begin{pmatrix} 0 & 1 \\ 1 & 0 \end{pmatrix}, \quad \gamma_i = \begin{pmatrix} 0 & -\sigma_i \\ \sigma_i & 0 \end{pmatrix},$$

where σ_i, $i = 1,2,3$ are the Pauli matrices.

The solutions of the Dirac equation provides a representation of the Lorentz group G. In fact for $g \in G$ we define

$$S(g) = \begin{pmatrix} g & 0 \\ 0 & g^{*-1} \end{pmatrix},$$

and one can then prove the following: If $u(x)$ is a solution of the Dirac rquation, then $u_g(x) = S(g)u(\Lambda_g^{-1}x)$ is also a solution, i.e. the Dirac equation is G covariant.

A solution of the Dirac equation has the following spectral representation

$$u(x) = (2\pi)^{-3/2} \sum_{s=1}^{2} \int \left[e^{-ik\cdot x} u_+(\underline{k},s) f_+(k,s) + e^{ik\cdot x} u_-(-\underline{k},s) f_-(k,s) \right] dk$$

where $k = (\omega_k, k_1, k_2, k_3) = (\omega_k, \underline{k})$, $\omega_k = (|\underline{k}|^2 + m^2)^{1/2}$, $k\cdot x = \omega_k x_0 - \underline{k}\cdot\underline{x}$, $dk = \frac{dk_1 dk_2 dk_3}{\omega_k}$ and $u_\pm(\underline{k},s)$ are the four eigenvactors of

$$D(k) = \gamma_0 \left(\sum_{i=1}^{3} k_i \gamma_i + m 1 \right),$$

i.e. $D(k)u_\pm(\underline{k},s) = \pm\,\omega_k u_\pm(\underline{k},s)$, normalized such that $u_\pm^* u_\pm = \omega_k$.

There is a natural Hilbert space associated with the data $f = \{f_+, f_-\}$ in fact let h_m denote the mass-hyperboloid

$$h_m = \{k = (k_0, \underline{k}) \in \mathbb{R}^4 \mid k_0 = \omega_k\} \quad .$$

It can be shown that dk is a G-invariant measure on h_m.

<u>Definition</u>. Let $H_\pm = L_2(h_m, dk, \mathbb{C}^2)$, i.e. $\mathbb{C}^2$-valued functions on h_m which are square integrable with respect to dk. The "one-particle" Hilbert space H_1 is given by $H_1 = H_+ \oplus H_-$.

This means that suitable data $f = \{f_+, f_-\}$ can be considered as elements in H_1. One can now show that the G-covariance of the Dirac equatiòn and the spectral representation for a solution together provides us with a unitary representation $g \to U_1(g)$ of G on H_1.

We can now in complete analogy with the Schrödinger case construct the anti-symmetric Fock-space $H = \bigoplus_{n=0}^{\infty} \Lambda^n H_1$ and the Fock-representation $f \to a(f)^*$ of the CAR. Furthermore we can write

$$a(f) = \sum_{s=1}^{2} \int [f_+(k,s)\alpha_+(k,s) + f_-(k,s)\alpha_-(k,s)]dk \ ,$$

where $\alpha_\pm(k,s)$ are densely defined operators in H_1.

The adjoints $\alpha_\pm(k,s)^*$ again only make sense as sesquilinear forms and the following anticommutation relations are fulfilled when properly interpreted

$$\alpha_\pm(k,s)\alpha_\pm(k',s')^* + \alpha_\pm(k',s')^*\alpha_\pm(k,s) = \omega_k \delta(\underline{k}-\underline{k}')\delta_{ss'} 1 \quad ,$$

with all other anti-commutators vanishing.

<u>The Dirac field</u> $\psi(x)$ is now defined by the following sesquilinear form

$$\psi(x) = (2\pi)^{-3/2} \sum_{s=1}^{2} \int \left[e^{-ik\cdot x} u_+(\underline{k},s)\alpha_+(k,s) + e^{ik\cdot x} u_-(-\underline{k},s)\alpha_-(k,s)^* \right] dk \ ,$$

i.e. $\psi(x)$ is a sesquilinear-form-valued solution of the Dirac equation. The associated energy-momentum operator $P = (P_0, \underline{P})$ is given by

$$P = \sum_{s=1}^{2} \int k[\alpha_+^* \alpha_+ + \alpha_-^* \alpha_-]dk \ ,$$

and each component P_μ defines a self-adjoint operator in the state-space H.

The unitary representation $g \to U_1(g)$ of G on H_1 lifts in complete analogy with the Schrödinger case to a unitary representation $g \to U(g)$ of G on H, and P transforms as a vector i.e.

$$U(g)\,P\,U(g)^{-1} = \Lambda_g P \ ,$$

i.e. the free mass-operator $P^2 = P_0^2 - P_1^2 - P_2^2 - P_3^2$ is invariant, $U(g)P^2U(g)^{-1} = P^2$.

The free mass-operator P^2 is the "relativistic" generalization of the square of the free Schrödinger mass operator M_0.

The Lorentz group G can be considered as a "relativistic" generalization of the homogeneous Galilei group with the important difference however that the Lorentz boosts do not commute, contrary to the Galilei boosts. There is in a certain sense a curvature.

One can now in analogy with the Schrödinger case introduce position operators as the generators of Lorentz boost and relative positive operators as generators of relative Lorentz boosts. That is, one can introduce the notion of relative distances even a theory which only is Lorentz invariant and not fully Poincaré invariant.

3.2. Lorentz covariant generalizations of H_S.

The covariant generalization of H_0 is given by the energy-momentum P. We shall now construct covariant generalizations of the Schrödinger interaction Hamiltonian $\frac{1}{2}\int :\nabla V \nabla V: dx$. We recall that the Coulomb field $V(x)$ is a solution of the Poisson equation $-\Delta u = p$ where p is of the form $\phi^*\phi$.

In the transition from electrostatics to Maxwell's theory one replaces the Poisson equation with the wave equation $\frac{\partial^2 u}{\partial t^2} - \Delta u = j$, where j is the electromagnetic current, transforming as a vector under Lorentz transformations. The Lorentz covariant generalization of $\frac{1}{2}\int |\nabla u|^2 dx$ will be chosen as the energy-momentum of the wave equation expressed as an integral over a light-cone.

Let us consider the retarded solution of the wave equation $\frac{\partial^2 u}{\partial t^2} - \Delta u = j(x,t)$, $x \in \mathbb{R}^3$, $t \in \mathbb{R}$,

$$u(x,t) = \frac{1}{4\pi}\int \frac{j(y,t-|x-y|)}{|x-y|}\,dy \equiv (D_R * j)(x,t) \ .$$

The energy associated with this solution can be expressed as an integral over $C_- = \{(x,t) \in \mathbb{R}^4 \mid t = -|x|\}$

$$p_0(u) = \frac{1}{2}\int_C \nabla u \overline{\nabla u}\,dx \ ,$$

where the gradients are tangential to C_- and we have surpressed the contraction over Lorentz indices. In fact the energy, associated with the wave equation, when expressed as an integral over C_-, looks similar to the electrostatic energy (compare [2]).

There are similar formulas for the momentum-components $p_i(u)$, $i = 1,2,3$,

It is now tempting to generalize the Schrödinger theory in the following way: The Coulomb field

$$V(x) = \frac{1}{4\pi} \int \frac{\rho(y)}{|x-y|} dy$$

with $\rho(x) = \sqrt{4\pi}\, e\, \phi(x)^* \phi(x)$ is replaced by

$$A(x,t) = (D_R * J)(x,t) ,$$

where the current J is sesquilinear in the Dirac field $\psi(x,t)$, i.e. $\sqrt{4\pi}\, e\, \psi^* \gamma_0 \gamma_\mu \psi$ or $\sqrt{4\pi}\, \kappa\, \psi^* \gamma_0 \psi$ for example. This field $A(x,t)$ is then inserted into the expression for the energy-momentum $p(\cdot)$ of the wave equation and this operator valued vector is added to the free energy-momentum P. The two choices for J above correspond to "electromagnetism" and "gravitation". Let us in the following consider the "electromagnetic" case, i.e.

$$J_u(x,t) = \sqrt{4\pi}\, e\, \psi(x,t)^* \gamma_0 \gamma_\mu \psi(x,t) ,$$

and in the following we surpress the index μ.

The associated "light-cone" energy is given by

$$(P_A)_0 = \frac{1}{2} \int_{C_-} :\nabla A \nabla A: dx ,$$

where the normal ordering is crucial, and there are similar formulas for the other components of P_A. Each component of P_A in fact defines a sesquilinear form. Let us now express P_A in terms of the $\alpha_\pm$'s, in fact

$$\begin{aligned} P_A = \lim_{\varepsilon \downarrow 0} \frac{1}{2} \int d\mu_1 \cdots d\mu_4 \Big[& p_\varepsilon(1,2;3,4) c_1^* c_3^* c_4 c_2 + p_\varepsilon(-1,-2;-3,-4) d_2^* d_4^* d_3 d_1 \\ & -2p_\varepsilon(1,2;-3,-4) c_1^* d_4^* d_3 c_2 + 2p_\varepsilon(1,-2;-3,4) c_1^* d_2^* d_3 c_4 \\ & + \Big[2p_\varepsilon(1,2;3,-4) d_4^* c_1^* c_3^* c_2 + 2p_\varepsilon(1,-2;-3,-4) c_1^* d_4^* d_2^* d_3 \\ & + p_\varepsilon(1,-2;3,-4) c_1^* d_2^* c_3^* d_4^* + \text{adjoint} \Big] \Big] , \end{aligned}$$

with $c_1 = \alpha_+(k_1,s_1)$, $d_1 = \alpha_-(k_1,s_1)$ etc., $\int d\mu = \Sigma_s \int dk$ and the kernels p_ε are given by

$$p_\varepsilon(1,2;3,4) = \frac{\alpha}{\pi}\,\bar{u}_1\gamma u_2\bar{u}_3\gamma u_4 \cdot \sigma(k_1-k_2-i\varepsilon, k_3-k_4-i\varepsilon)$$

$$\sigma(z_1,z_2) = -\frac{2i}{\pi^2}\,\frac{1}{((z_1+z_2)^2)^2}\left[\frac{z_1}{z_1^2}+\frac{z_2}{z_2^2}\right],$$

where $\alpha = \frac{e^2}{4\pi}$, $u_j = u_+(\underline{k}_j, s_j)$, $u_j = u_-(-\underline{k}_j, s_j)$ and ε is understood as the time-like vector $(\varepsilon, \underline{0})$. Furthermore we have put $\bar{u}_j = u_j\gamma_0$.

We note that σ behaves as a vector under Lorentz transformations.

It can be shown that this formula for P_A defines a densely defined sesquilinear form which in the "non-relativistic" limit reduces to the corresponding Schrödinger interaction Hamiltonian.

We conjecture that $(P + P_A)_0$ is bounded from below. In the non-relativistic limit it follows from results of Lieb and Thirring on stability of matter.

3.3. The mass operator and relative dymanics.

It is now tempting to consider the Lorentz invariant mass "operator" $M^2 = (P + P_A)^2$, which if it defines a self-adjoint operator, can be used to generate proper time dynamics.

We shall assume that the Schrödinger mass operator M_S and M_S^2 generate the same "physics" i.e. scattering operators etc.

The Lorentz invariant mass operator M^2 will in analogy with the Schrödinger case generate the internal or relative dynamics.

The study of $P + P_A$ and M^2 is a big program and it is natural to start with a study of the restriction of $P + P_A$ to the two-particle sector for example, and do the spectral and scattering theory. This is work in progress.

There are many possible experimental tests of a putative theory of this sort, and will be suggested elsewhere. In fact the conservation laws are different in this approach but reduce to the conventional ones in the "non-relativistic limit".

4. Conclusions.

We have suggested a new type of "relativistic" quantum theory, which is Lorentz invariant but not fully Poincaré invariant. The theory have the correct non-relativistic limit. Preliminary investigations indicates that there are no "ultra-violet" or "infrared" divergencies in the theory.

One can introduce relative position operators in terms of the generators of relative Lorentz boosts.

This work is just the start of a long term study of Lorentz invariant quantum theories.

References.

[1] Pauli, W.: Einige die Quantenmechanik betreffenden Erkundingsfragen. Zeits. f. Physik 80, 513 (1933).

[2] Lundberg, L.-E.: The Klein-Gordon Equation with Light-Cone Data. Comm. Math. Phys. 62, 107 (1978).

A CHARACTERIZATION OF
DILATION-ANALYTIC OPERATORS

T. Paul

Centre de Physique Théorique,
Marseille.

1. Introduction

The purpose of this talk is to present a criterion of Schrödinger operators which is the result of a joint work with E. Balslev and A. Grossman [1].

The concept of dilation analyticity has proved very fruitful in the study of Schrödinger operators ([2],[3]).

Given a unitary representation of the dilation group $\mathbb{R}^+$ by the family of operators $\{U(p);\ p > 0\}$ acting on a Hilbert space $\mathcal{H}$, dilation analyticity of an operator V on $\mathcal{H}$ is defined by requiring that the operator valued function $V(p) = U(p)\ V\ U(p)^{-1}$ have an analytic extension in p to a given sector containing the half-line.

From the beginning of the theory the problem of characterising the class of dilation-analytic operators has been investigated. Althought positive answer has been given for local potentials (i.e. operators of multiplication by a given function) in [4], only bounded integral operators, whose kernels are analytic in both variables have been treated [5]. The choice of the representation space seems to be very important for this problem : indeed, although dilation-analytic local potentials are characterized by simple analyticity conditions, the same operators in momentum spaces are usualy given by very singular kernels. An example is the Coulomb potentials $V(q)=-\frac{1}{q}$ which gives rise in three dimensions to the integral operator in momentum space given by a kernel proportional to $(\vec{p}-\vec{p}\,')^{-2}$.

This example shows that the use of a representation space in which a large class of operators could be represented as integral operators with analytic kernels could be very useful for a general investigation of this problem. The representation of quantum mechanics on a space $\mathcal{H}_\alpha$ of analytic functions on a half-plane developped in [6], [7] seems particularly well aclapted.

If we write the momentum space of quantum mechanics as $L^2(\mathbb{R}^+, h; p^{N-1}\, dp)$ (with for example $h = L^2(S^{n-1})$), the space we use is a space $\mathcal{H}_\alpha^h = \mathcal{H}_\alpha \otimes h$ of h-valued analytic functions on the half-plane, square integrable with respect to a two-dimensional measure $d\mu_\alpha(z)$ defined below.

The correspondance between the two spaces is explicitly given by associating to each function $\psi \in L^2(\mathbb{R}^+, h; p^{N-1}\, dp)$ the function f on $\mathbb{C}^+$ given by

$$f(z) = \frac{1}{\sqrt{\pi\,\Gamma(\beta + \frac{N}{2} - 1)}} \int_0^{+\infty} p^\beta e^{iz\frac{p^2}{2}} \psi(p)\, p^{N-1}\, dp \qquad (1.1)$$

(see [1]).

This representation has the advantage first of all, that analyticity is already built into the theory. Secondly, a large class, in particular all H_0- bounded operators are represented by an integral kernel, analytic in both variables, in a sense given below. Thirdly the space $\mathcal{H}_\alpha^h$ has a (generalized) reproducing kernel. In this paper, we analyze the class of dilation analytic vectors and operators in this representation.

In §2 we introduce the Hilbert spaces used in this paper. §3 is devoted to the study of dilation analytic vectors and §4 to dilation analytic-operators. The reader is refered to [1] for the proofs of the statements.

2. Hilbert spaces

a) The space $\mathcal{H}_\alpha$

let $\alpha > -1$ be a real number (kept fixed). We denote by $\mathcal{H}_\alpha$ the space of analytic functions f on $\mathbb{C}^+ = \{z = x + iy,\ y > 0\}$ such that

$$(f,f)_\alpha \equiv \iint_{\mathbb{C}^+} |f(z)|^2 y^\alpha \, dx\, dy = \int_{\mathbb{C}^+} |f(z)|^2 \, d\mu_\alpha(z) < +\infty \tag{2.1}$$

Equipped with the scalar product

$$(f,g)_\alpha = \int_{\mathbb{C}^+} \overline{f(z)} \, g(z) \, d\mu_\alpha(z) \tag{2.2}$$

$\mathcal{H}_\alpha$ is a Hilbert space with reproducing kernel defined as follows :

for $z \in \mathbb{C}^+$ and $w \in \mathbb{C}^+$, consider the function

$$q_z(w) = \frac{\alpha+1}{4\pi} \left(\frac{w - \bar{z}}{2i} \right)^{-\alpha-2} \tag{2.3}$$

Then

i) q_z belongs to $\mathcal{H}_\alpha$ for every $z \in \mathbb{C}^+$

ii) $(q_z, f) = f(z)$ for every f in $\mathcal{H}_\alpha$ (2.4)

The norm of q_z in $\mathcal{H}_\alpha$ is

$$\| q_z \|_\alpha = \sqrt{\frac{\alpha+1}{4\pi}} \, (\mathrm{Im}\, z)^{-\frac{\alpha}{2}-1} \tag{2.5}$$

By Schwarz inequality we deduce from (2.4) and (2.5) that, for every f in $\mathcal{H}_\alpha$,

$$|f(z)| \leq \sqrt{\frac{\alpha+1}{4\pi}} \, (\mathrm{Im}\, z)^{-\frac{\alpha}{2}-1} \, \| f \|_\alpha \tag{2.6}$$

The link (1.1) between $\mathcal{H}_\alpha$ and $L^2(\mathbb{R}^+, p^{N-1}\, dp)$ has been investigated in [1] and [6].

b) The space $\mathcal{H}_\alpha^h$

As we have seen in the introduction the space $\mathcal{H}_\alpha$ is used to describe the "radial" behaviour of a one-body hamiltonian in n dimensions. In order to describe the full hamiltonian we use, as is customary [6], a space of functions on $\mathbb{C}^+$ with values in the Hilbert space $h = L^2(S^{n-1})$ of square

integrable functions on the unitsphere in $\mathbb{R}^n$. The precise nature of h and of the Hilbert space with reproducing kernel are not important here; however to fix the ideas we treat the case $\mathcal{H}_\alpha$.
We denote by $\mathcal{H}_\alpha^h$ the set of all h-valued functions f, analytic on $\mathbb{C}^+$ such that

$$(f,f)_{\alpha,h} \equiv \iint |f(z)|_h^2 \, d\mu_\alpha(z) < +\infty \tag{2.7}$$

when $|.|_h$ denotes the norm in h. The $\mathcal{H}_\alpha^h$ is a Hilbert space which, in general does not have a reproducing kernel.
However, for each $\chi \in h$ and $z \in \mathbb{C}^+$, we may consider in $\mathcal{H}_\alpha^h$ the vector q_z^χ defined by

$$q_z^\chi(z') = q_z(z') \cdot \chi \tag{2.8}$$

The we have [1]:

$$(q_z^\chi, f)_{\alpha,h} = (\chi, f(z))_h \tag{2.9}$$

which can be seen as a weak reproducing property.
Moreover we have, [1],

Proposition 1 :

If f belongs to $\mathcal{H}_\alpha^h$, then, for every $z \in \mathbb{C}^+$, one has

$$|f(z)|_h \leq \sqrt{\frac{\alpha+1}{4\pi}} (\operatorname{Im} z)^{-1-\frac{\alpha}{2}} \| f \|_{\alpha,h} \tag{2.10}$$

We finish this section by introducing the following natural unitary representation of the one-parameter dilation group

$$(U(k)f)(z) = k^{1+\frac{\alpha}{2}} f(kz) \quad (k>0,\ z \in \mathbb{C}^+) \tag{2.11}$$

We shall be concerned with complexifications of this representation and start by defining the following domains :

let a be such that $0 \leq a \leq \frac{\pi}{2}$, we denote S_a the set :

$$S_a = \{ k \in \mathbb{C},\ -a < \arg(k) < a \} \tag{2.12}$$

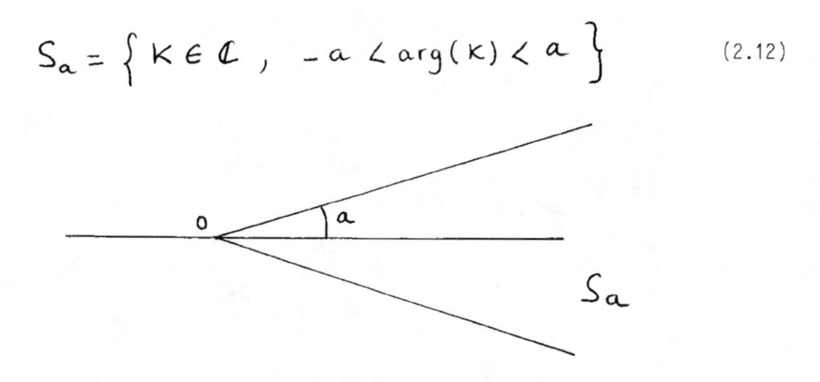

3. A. Characterisation of dilation-analytic vectors

We first recall the definition of S_a dilation analytic vector :
a vector $f \in \mathcal{H}^h_\alpha$ is S_a-dilation analytic if the family of vectors

$$f^k \equiv U(k) f \qquad (k>0) \tag{3.1}$$

in $\mathcal{H}^h_\alpha$ is the restriction, to the positive real axis, of an analytic $\mathcal{H}^h_\alpha$-valued function f^k defined for $k \in S_a$. This is the standard definition of dilation analyticity (see [2], [3], [4]) transported to the space $\mathcal{H}^h_\alpha$. It is stated in terms of the whole family f^k : so it is natural to ask for a criterion involving only f.

We will see that such a criterion will be given in the space $\mathcal{H}^h_\alpha$ in terms of analytic continuation properties of f into a suitable "cut pie" extending into the lower half-plane.

Let us define this "cut pie" :

for $0 \leq a \leq \frac{\pi}{2}$ we denote τ_a the set

$$\tau_a = \{ z \in \mathbb{C} \;/\; -a < \arg(z) < \pi + a \} \tag{3.2}$$

$$= e^{-ia} (\mathbb{C}^+)^{1 + \frac{2a}{\pi}}$$

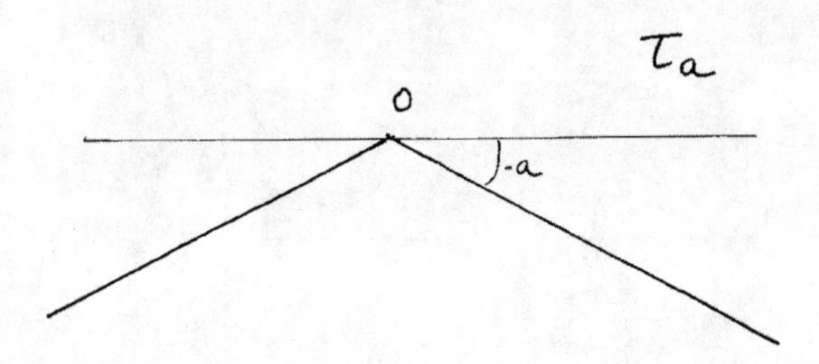

We now define $\mathcal{D}(a,\alpha)$ as the set of all f in $\mathcal{H}^h_\alpha$ which are such that

i) the h-valued function $z \to f(z)$ $(z \in \mathbb{C}^+)$ is the restriction to $\mathbb{C}^+$ of an h-valued function defined and analytic in τ_a.

ii) for every ϕ such that $-a < \phi < a$, the h-valued function $z \to f(e^{-i\phi} z)$ is square-integrable over $\mathbb{C}^+$ with respect to the measure $y^\alpha \, dx \, dy$ (i.e. belongs to $\mathcal{H}^h_\alpha$) :

$$N\phi = \iint | f(e^{-i\phi} z) |^2_h \, y^\alpha \, dx \, dy < +\infty \tag{3.3}$$

iii) the numbers $N\phi$ are uniform ly bounded in any closed interval contained in the open interval $(-a,a)$: for every $f > 0$, these exists a C_δ such that $N_\phi \leq C_\delta$ for every ϕ in $[-a + \delta, a - \delta]$.

The main result of this section is the following :

Proposition 2 :

$\mathcal{D}(a,\alpha)$ is the set of S_a- dilation analytic vectors.

The proof of this proposition can be found in [1] and uses extensively the reproducing kernel properties.

As a direct application of this proposition it is possible to recover the result concerning the characterisation of dilation-analytic vectors in $L^2(\mathbb{R}^n)$ given in [4] . (See again [1]).

3. A characterisation of dilation-analytic operators

This section is devoted to the study of operators in $\mathcal{H}_\alpha^h$.

We first see that, for a large class of self-adjoint operators H_0^h in $\mathcal{H}_\alpha^h$, the set of H_0^h-bounded operators in $\mathcal{H}_\alpha^h$ can be represented by integral operators (in a sense given below) with analytic kernels. Then dilation analyticity can be expressed as properties of analytic continuation of the kernels in suitable domains.

Let H_0^h be a self-adjoint operator in $\mathcal{H}_\alpha^h = \mathcal{H}_\alpha \otimes h$; assume that H_0^h is of the form $H_0 \otimes \mathbb{1}_h$ where H_0 acts in $\mathcal{H}_\alpha$ ("radial" property). Assume furthermore that the domain of H_0 contains all the functions q_z^χ $(z \in \mathbb{C}^+, \chi \in h)$.

Let V be symmetric and H_0^h -bounded i.e. the domain of V contains the domain of H_0^h and V is bounded as an operator from the domain of H_0^h with the graph norm into $\mathcal{H}_\alpha^h$. Since $V \subseteq V^*$, V is closable. By the assumption on V, the vector Vq_z^χ is in $\mathcal{H}_\alpha^h$ for every $z \in \mathbb{C}^+$, $\chi \in h$. For every $z' \in \mathbb{C}^+$, consider the vector in h defined as $(Vq_z^\chi)(z')$. Define $v(z,\bar{z}')\chi$ for $\chi \in h$ and $z,z' \in \mathbb{C}^+$ by

$$v(z,\bar{z}')\chi = (V q_{z'}^{\chi})(z) \tag{4.1}$$

The we have the following.

Proposition 3 [1] :

- for fixed z,z', the correspondance $\chi \to v(z,z')\chi$ is a bounded linear operator in h and $v(z,\bar{z}')^* = v(z',\bar{z})$
- for every f in the domain of V and every $\chi \in h$,

$$(\chi, (Vf)(z))_h = \int (\chi, v(z,\bar{z}') f(z'))_h \, d\mu_\alpha(z') \qquad (4.2)$$

where the integral on the r.h.s. is absolutely convergent.

Remark : The equality (4.2) can be written

$$(Vf)(z) = \int v(z,\bar{z}') f(z') \, d\mu_\alpha(z')$$

where the integral in the r.h.s. is weakly convergent.

We now give the definition of dilation analyticity for operators (see[2]).

Definition

V is S_a-dilation analytic if the map $K \to U(K)\, V\, U(K)^{-1}$ from $\mathbb{R}^+$ into $\mathcal{B} = \mathcal{B}(\mathcal{D}(H_0), \mathcal{H}^h_\alpha)$ (where $\mathcal{D}(H_0^h)$ is the domain of H_0^h equipped with the graph norm) has an analytic extension V^K to S_a.

We define also F_a as the subject of $\mathbb{C}^2$ defined as follows :

$$F_a = \bigcup_{-a<\varphi<a} (e^{i\varphi}\mathbb{C}^+) \times (e^{i\varphi}\mathbb{C}^-)$$

where $\mathbb{C}^-$ is the lower open half-plane. F_a can also be written as :

$$F_a = \Big\{ (z,w) \in \mathbb{C}^2 / z \in \mathcal{T}_a \text{ and } \operatorname{Max}(-\pi - a, (\arg z) - 2\pi) < \arg \bar{w} < \operatorname{Min}(\arg z, a) \Big\}$$

We can now state our main result.

Proposition 4 [1]

Suppose that H_0^h and V are as above. Then V is S_a-dilation analytic if and only if the three conditions below hold :

i) The family $v(z,\bar{w})$ has a $\mathcal{B}(h)$-valued analytic continuation from $\mathbb{C}^+ \times \mathbb{C}^-$ to F_a

ii) For each $K \in S_a$, the operator U^K, defined by the kernel

$$v^K(z,\bar{z}') = K^{d+2}\, v(Kz, K\bar{z}')$$

is H_0^h-bounded

iii) the operator norms

$$\|(H_0^h - i)^{-1} V^{\exp(-i\varphi)}\|$$

in the space $\mathcal{H}_\alpha^h$, are uniformely bounded in any closed interval $-a + \delta \leq \phi \leq a + \delta$

Proof : [1]

a) assume that V is S_a-dilation analytic for $k > 0$, the operator $V^K = U(K)\, V\, U(K)^{-1}$ is H_0^h-bounded.

By proposition 3 a bounded operator in h is defined by

$$v^K(z,\bar{z}')\,\chi = v^K q_{z'}^{\chi}(z)\,.$$

One has

$$(\chi, v^K(z,\bar{z}')\chi')_h = \left(q_z^{\chi}, U(K)\, V\, U(K)^{-1} q_{z'}^{\chi'}\right)_{\mathcal{H}_\alpha^h}$$

$$= K^{d+2}\left(\varphi_{Kz}^{\chi}, V q_{Kz'}^{\chi'}\right)_{\mathcal{H}_\alpha^h}$$

so

$$v^K(z,\bar{z}') = K^{d+2}\, v(Kz, K\bar{z}')$$

By the assumption that V is S_a-dilation analytic, $(q_z^{\chi}, V^k q_{z'}^{\chi'})$ has, for fixed z,z' in $\mathbb{C}^+$, an analytic continuation into $K \in S_a$.

This gives the analyticity of $v(z,\bar{z}')$ in F_a. (ii) and (iii) follow from the definition (analyticity and continuity).

b) assume the family $v(z,\bar{z}')$ of operators in $\mathcal{B}(h)$ satisfies (i), (ii) and (iii).

Define $v^k(z,\bar{z}')$ by

$$v^k(z,\bar{z}') = k^{\alpha+2} v(kz, k\bar{z}')$$ and

V^k by the "kernel" $v^k(z,\bar{z}')$. Then, $(q_{z'}^{\chi'}, V^k q_z^k) = (\chi', v^k(z,\bar{z}')\chi)_h$ is analytic in z and z'. From the density of the linear space of $\{q_z^\chi\}$ and (iii) we get the analyticity of V^k.

Remark : as far as operators are concerned, it is of course not possible to characterize the kernels of bounded integral operators. Based on well-known criterion for an integral operator to be bounded it is possible to give a sufficient condition for an integral operator in momentum space to be dilation-analytic (see [1]).

Acknowledgements :

I would like to thank the "Departement de Physique Théorique et Mathématique, U.C.L., Louvain-la-Neuve, Belgique" where the final draft of the manuscript was performed.

I also acknowledge the "Matematisk Institut, Aarhus Universitet" for kind hospitality during the conference.

References :

1. E. Balslev, A. Grossmann and T. Paul, A characteristisation of dilation-analytic operators (to appear in "Annales de l'Institut H. Poincaré").

2. J. Aguilar and J.M. Combes, A class of analytic perturbations for one-body Schrödinger hamiltonians - Commun. Math. Phys. 22, 269-279 (1971)

3. E. Balslev and J.M. Combes, Spectral properties of many-body Schrödinger operators with dilation-analytic interactions. Commun. Math. Phys. 22, 280-299 (1971)

4. D. Babbitt and E. Balslev, A characterisation of dilation-analytic potentials and vectors. J. Funct. Analysis 18, 1-14 (1975)

5. A. Dionisi Vici, A characterisation of dilation analytic integral kernels, Lett. Math. Phys. 3, 533-541 (1979)

6. T. Paul, Functions analytic on the half-plane as quantum mechanical states. J. Math. Phys. 25, 3252-3263 (1984)

7. T. Paul, Affine coherent states for the radial Schrödinger equation I. Radial harmonic oscillator and hydrogen atom. Preprint CPT 84/P. 1710, Marseille.

ASYMPTOTIC AND APPROXIMATE FORMULAS
IN THE INVERSE SCATTERING PROBLEM FOR THE SCHRÖDINGER OPERATOR

Yoshimi Saitō
Department of Mathematics
The University of Alabama at Birmingham
Birmingham, AL 35294 U. S. A.

§1. Introduction

In this article we are going to review some new results (Saitō [22, 23, 24, 26]) in the inverse scattering problem for the Schrödinger operator.

Let us consider the differential expression

$$h = -\Delta + Q(x) \qquad (x \in \mathbb{R}^3) \tag{1.1}$$

where Δ is the Laplacian and $Q(x)$ is a real-valued, measurable function on $\mathbb{R}^3$ such that we have

$$|Q(x)| \le C_0(1 + |x|)^{-\beta} \qquad (x \in \mathbb{R}^3) \tag{1.2}$$

with positive constants C_0 and β satisfying

$$\beta > 1, \tag{1.3}$$

i.e., $Q(x)$ is a short-range potential. H and H_0 denote unique self-adjoint extensions of h and $h_0 = -\Delta$ restricted to $C_0^\infty(\mathbb{R}^3)$ in $L_2(\mathbb{R}^3)$, respectively. Since the multiplication operator $Q(x)\times$ is a bounded linear operator on $L_2(\mathbb{R}^3)$, the domain $D(H)$ of H coincides with the domain $D(H_0)$ of H_0 and they are equal to $H^2(\mathbb{R}^3)$, the Sobolev space of the second order on $\mathbb{R}^3$ (see, e.g., Kato [10], Chapter 5).

The scattering theory for the Schrödinger operator H is developed as follows (see e.g., Agmon [1], Amrein et al. [3], Saitō [20]): The wave operators $W_\pm$ are defined by

$$W_\pm = s - \lim_{t\to\pm\infty} e^{itH} e^{-itH_0} \quad \text{in} \quad L_2(\mathbb{R}^3) \tag{1.4}$$

(Kuroda [11]). By the use of $W_\pm$ the scattering operator S is defined by

$$S = W_+^* W_-, \tag{1.5}$$

where W_+^* is the adjoint of W_+ in $L_2(\mathbb{R}^3)$. S is a unitary operator on $L_2(\mathbb{R}^3)$. Further, it is known that there exists a family $\{S(k)\}_{k>0}$ of unitary operators on $L_2(S^2)$ such that

$$(FSF^*G)(\xi) = \{S(|\xi|)G(|\xi|\cdot)\}(\tilde{\xi}) \tag{1.6}$$
$$\left(G \in C_0^\infty(\mathbb{R}^3),\quad \xi \in \mathbb{R}^3,\quad \tilde{\xi} = \xi/|\xi|\right),$$

where S^2 is the unit sphere in $\mathbb{R}^3$, F and F^* are the usual Fourier transform and the inverse Fourier transform (Agmon [1], Saitō [20,21]). $S(k)$ is called the S-matrix in physics. Set

$$F(k) = -2\pi ik^{-1}\left(S(k) - I\right) \tag{1.7}$$

with the identity operator I on $L_2(S^2)$. $F(k)$ is a Hilbert-Schmidt operator with integral kernel $F(k,\omega,\omega')$, $k > 0$, ω, $\omega' \in S^2$ if $Q(x)$ satisfies (1.2) with $\beta > 2$. $F(k,\omega,\omega')$ is called the scattering amplitude (see, e.g., Amrein et al [3]).

In the inverse scattering problem for H, roughly speaking, we are going to reconstruct the potential $Q(x)$ from given the scattering data $S(k)$ (or $F(k)$). Especially we are interested in seeking "practical" methods of recovering the potential $Q(x)$ from the scattering data. In quantum physics the scattering data $F(k)$ are measured by experiments, where we can get $F(k)$ for only a finite number of k.

The core of our results consists of two formulas: the asymptotic formula enables us to recover the potential as a high-energy limit of the scattering data when the energy value k goes to ∞ (see §3). Using the approximation formula, we can recover the potential to any degree of accuracy if we know only the scattering data for some sufficiently high energy value k, so that the scattering data for k around $k = \infty$ are not necessary in order to get a good approximate value for the potential (see §4).

Our asymptotic formula holds for general short-range potentials so that we can prove the uniqueness of the inverse scattering problem and give a reconstruction formula for quite a wide class of potentials. We shall also be able to discuss the characterization problem. Our approximation formula will give us a practical way to approximate the potential using the scattering data which can be obtained by experiment.

We shall review some known results on the inverse scattering problem for the Schrödinger operator in §2. In §3 we shall state an asymptotic behavior of $F(k)$ and discuss the implication of it. An approximate formula will be given in §4 in order to recover the potential

$Q(x)$ with any accuracy when we know only the scattering data for some sufficiently high energy value k. Finally, in §5, we shall discuss some related problems which might be interesting.

§2. Some known results

The inverse scattering problem for the one-dimensional Schrödinger operator

$$H_1 = -\frac{d^2}{dr^2} + q(r) \qquad (r \in (0,\infty)) \tag{2.1}$$

with a boundary condition at $r = 0$ has been studied from around 1950. The classical works of Gel'fand-Levitan [8] and Agranovich-Marchenko [2] should be mentioned. In their works, roughly speaking, the potential was recovered by solving a Fredholm integral equation whose kernel is constructed by the use of the Fourier transform of the spectral measure or the scattering data.

After many efforts to extend the results of the Schrödinger operator H_1 in $\mathbb{R}^1$ to the Schrödinger operator

$$H = H_3 = -\Delta + Q(x) \qquad (x \in \mathbb{R}^3) \tag{2.2}$$

in $\mathbb{R}^3$, Newton [16-18] produced the first successful works in this direction. His method was applied to the Schrödinger operator H_2 in R^2 by Cheney [5].

In all these works we need the spectral measure on the whole real line or the scattering data for all energy numbers $k \geq 0$ to recover the potential, and the potentials are assumed to decrease to zero near infinity more rapidly than short-range potentials. In fact Newton [16] gives a sufficient condition that $Q(x) = 0(|x|^{-3-\varepsilon})$ and $|\nabla Q(x)| = 0(|x|^{-2-\varepsilon})$ with $\varepsilon > 0$ at infinity.

There is another method for recovering the potential which may be called a high energy limit method. Let $\beta > 3$ in (1.2) and let $F(k)$ be as in (1.7). Then the potential $Q(x)$ is in $L_1(\mathbb{R}^3)$ and the scattering amplitude $F(k,\omega,\omega')$ exists. Faddeev [6] showed the formula

$$\lim_{\substack{k\to\infty \\ \xi=k(\omega-\omega')}} F(k,\omega,\omega') = \frac{1}{4\pi}\int_{\mathbb{R}^3} e^{-i\xi x} Q(x)\,dx \qquad (\xi \in \mathbb{R}^3) \tag{2.3}$$

under the above assumption. Here the limit is taken so that k goes to ∞ keeping the relation $\xi = k(\omega - \omega')$ for a given $\xi \in \mathbb{R}^3$. The potential $Q(x)$ can be recovered by the usual Fourier inversion formula

As for the extensions of Faddeev's result, see Newton [16], Lemma 3.1, and Saitō [22], Theorem 5.1. In the work of Faddeev [6], we don't need any low energy scattering data but we need the scattering data around $k = \infty$ in order to recover the potential. Another remark on the Faddeev work is that the limit is taken in (2.3) so that the difference between ω and ω' becomes smaller and smaller. This means that the measurement of the scattering data should be done around the singularities of $F(k,\omega,\omega')$ which makes the physical measurement very difficult (cf. a remark on Lemma 3.1 of Newton [16], the 2nd column of p. 1698).

§3. Asymptotic behavior of the S-matrix

Let $F(k)$ be as in (1.7) and let us set

$$f(x,k) = k^2(F(k)\psi_{x,k},\psi_{x,k})_{L_2(S^2)}, \tag{3.1}$$

where

$$\psi_{x,k} = \psi_{x,k}(\omega) = e^{-ikx\omega} \in L_2(S^2) \qquad (x \in \mathbb{R}^3,\ k > 0,\ \omega \in S^2) \tag{3.2}$$

with the inner product $x\omega$ in $\mathbb{R}^3$. We regard $\psi_{x,k}$ as an element of $L_2(S^2)$ with parameters x and k. Saitō [22] showed the following asymptotic formula:

<u>Theorem 3.1</u>. Assume that $Q(x)$ is a general short-range potential, i.e., assume that $\beta > 1$ in (1.2). Let $f(x,k)$ be as in (3.1). Then there exists the limit

$$\lim_{k\to\infty} f(x,k) = -2\pi \int_{\mathbb{R}^3} \frac{Q(y)}{|y-x|^2}\,dy. \tag{3.3}$$

For the proof of this theorem see Saitō [22]. Let us discuss the implications of the formula (3.3). Fortunately the right-hand side of (3.3) takes a mathematically simple form. In fact this is a convolution of $Q(y)$ and $|y|^{-2}$. Thus, setting

$$f(x,\infty) = \lim_{k\to\infty} f(x,k), \tag{3.4}$$

we can rewrite (3.3) as

$$f(x,\infty) = -2\pi\{Q * |y|^{-2}\}(x). \tag{3.5}$$

By taking the Fourier transforms of the both sides "formally", we get

$$\{Ff(\cdot,\infty)\}(\xi) = -4\pi^3(FQ)(\xi)\cdot|\xi|^{-1} \qquad (\xi \in \mathbb{R}^3) \tag{3.6}$$

where we used the well-known formula

$$F(|y|^{-2})(\xi) = 2\pi^2|\xi|^{-1}. \tag{3.7}$$

Since the potential $Q(x)$ is assumed to be a general short-range potential, the Fourier transform FQ exists only in the sense of $S'(\mathbb{R}^3_\xi)$, the dual space of the space $S(\mathbb{R}^3_\xi)$ of rapidly decreasing functions. Therefore (3.6) should be regarded as a relation in $S'(\mathbb{R}^3_\xi)$. As is well-known, we cannot generally make the product of two distributions. But, if we can multiply the both sides of (3.6) by $|\xi|$, then we get

$$(FQ)(\xi) = -(4\pi^3)^{-1}|\xi|\{Ff(\cdot,\infty)\}(\xi). \tag{3.8}$$

It was shown in [23] that the above formal procedure could be justified rigorously as follows:

Lemma 3.2. (i) Let $f(x)$ satisfy

$$|f(x)| \leq C_1(1 + |x|)^{-(\beta-1)} \qquad (x \in \mathbb{R}^3) \tag{3.9}$$

with constants $C_1 > 0$ and $\beta > 1$. Then $|\xi|(Ff)(\xi)$ is well-defined as an element of $S'(\mathbb{R}^3_\xi)$.

(ii) A necessary and sufficient condition on $f(x)$ that the integral equation

$$f(x) = -2\pi \int_{\mathbb{R}^3} \frac{g(x)}{|x - y|^2}\, dx \tag{3.10}$$

has a unique real-valued solution $g(x)$ satisfying

$$|g(x)| \leq C_2(1 + |x|)^{-\beta} \qquad (x \in \mathbb{R}^3) \tag{3.11}$$

with constants $C_2 > 0$ and $\beta > 1$ is that $f(x)$ satisfies

$$|f(x)| \leq C_3(1 + |x|)^{-(\beta-1)} \qquad (x \in \mathbb{R}^3) \tag{3.12}$$

with a constant $C_3 > 0$ and that $F^*\{|\xi|(Ff)\}(x)$ is a real-valued function with the estimate

$$|F^*\{|\xi|(Ff)\}(x)| \leq C_4(1 + |x|)^{-\beta} \qquad (x \in \mathbb{R}^3) \tag{3.13}$$

where $C_3 > 0$ is a constant. Then the (unique) solution is expressed as

$$g(x) = (-4\pi^3)^{-1} F^*\{|\xi|(Ff)\}(x). \tag{3.14}$$

It follows from Lemma 3.2 that the potential $Q(x)$ is reconstructed by the formula (3.8). If $f(x) = 0$, then $g(x) = 0$ is a unique solution of the integral equation (3.10), which implies the uniqueness of the inverse scattering problem for short-range scattering, i.e., one and only one potential $Q(x)$ corresponds to a given S-matrix $S(k)$, $k > 0$. Further, using the above results, we can characterize a family $\{S_0(k)\}_{k>0}$ of unitary operators on $L_2(S^2)$ which is the S-matrix of a short-range scattering. In fact, let $F_0(k) = -2\pi ik^{-1}(S_0(k) - I)$ and define $f_0(x,k)$ by (3.1) with $F_0(k)$ replaced by $F(k)$. In order that $S_0(k)$ is an S-matrix, $f_0(x,k)$ should have the limit $f_0(x,\infty) = \lim_{k\to\infty} f_0(x,k)$ and $f_0(x,\infty)$ should satisfy the estimates (3.12) and (3.13) with $f_0(x,\infty)$ replaced by $f(x)$. Let $Q(x)$ be the potential given by (3.8) with $f_0(x,\infty)$ replaced by $f(x,\infty)$ and let $S(k)$, $k > 0$, be the S-matrix associated with $Q(x)$. If $S_0(k) = S(k)$ for all $k > 0$, then the family $\{S_0(k)\}_{k>0}$ is S-matrix associated with the short-range potential $Q(x)$. On the other hand if $S(k) \neq S_0(k)$ for some $k > 0$ then $S_0(k)$, $k > 0$, cannot be an S-matrix.

All these results were extended to the Schrödinger operator in $\mathbb{R}^N$ with $N \geq 2$ ([24]).

§4. Approximation formula

The asymptotic behavior which we gave in the preceding section can be said to be a kind of modification of the formula (2.3) by Faddeev [6]. In fact, both of them come from the idea of the Born approximation. Therefore though our formula (3.3) has shared its good point of simplicity with the Faddeev formula (2.3), it also shares a bad point that the measurement around the singularities becomes much more influential in the process of getting the limit as $k \to \infty$ in (3.3), and hence it is experimentally very hard to get the value of $f(x,k)$ around $k = \infty$. At the same time, since there are always some errors in the scattering data which are obtained by experiment, we cannot determine "exactly" the potential. Thus it would be interesting to construct a theory which guarantees that we can get a good approximate value for the potential if we get reasonably good scattering data.

Let $Q(x)$ satisfy (1.2) with $\beta > 1$ and let us define $Q_k(x)$, $k > 0$, by

$$Q_k(x) = -(4\pi^3)^{-1}F^*\{|\xi|(Ff(\cdot,k))\}(x). \tag{4.1}$$

Our approximation formula is stated as follows:

<u>Theorem 4.1.</u> Let $Q(x)$ satisfy (1.2) and

$$|D^\alpha Q(x)| \leq C_0(1 + |x|)^{-\rho} \qquad (x \in \mathbb{R}^3,\ |\alpha| = 1,2) \tag{4.2}$$

with

$$\frac{7}{4} < \beta \leq 2 \quad \text{and} \quad \frac{5}{2} < \rho \leq 3. \tag{4.3}$$

Here $\alpha = (\alpha_1, \alpha_2, \alpha_3)$ with nonnegative integer $\alpha_1, \alpha_2, \alpha_3$ is a multi-index, $|\alpha| = \alpha_1 + \alpha_2 + \alpha_3$ and

$$D^{\alpha} = D_1^{\alpha_1} D_2^{\alpha_2} D_3^{\alpha_3} \qquad \left(D_j = \frac{\partial}{\partial x_j}, \ x = (x_1, x_2, x_3)\right). \tag{4.4}$$

Then $Q_k(x)$ given in (4.1) is well-defined as an element of $L_2(\mathbb{R}^3)$ for each $k > 0$ and we have

$$\| Q - Q_k \|_{L_2(\mathbb{R}^3)} \leq \frac{C}{k} \qquad (k \geq a) \tag{4.5}$$

where a is any positive number and C is a constant depending only on $Q(x)$ and a.

For the proof see Saitō [26]. It is also shown that C depends only on the constant C_0 in (1.2) and (4.2) if a is taken sufficiently large and that $C = C(C_0)$ is bounded when C_0 moves in a bounded set in $[0, \infty)$. Therefore, when we know the range of the "size" C_0 of $Q(x)$ in advance, we can approximate the potential Q with any accuracy by choosing a sufficiently high energy number k in (4.5).

It seems that there are few works which discuss the error estimates in the inverse scattering problem, while Prosser [19] gave an error estimate in the case of Born approximation under much more restrictive conditions on the potential $Q(x)$.

§5. Some problems in the inverse scattering problems for the Schrödinger operator.

Let us mention several problems related to the asymptotic formula and approximation formula given in §3 and §4.

(1) Different types of approximation formulas. In §4 an approximation formula of L_2-type was given. Some other types of the approximation formulas would be useful. Especially a pointwise approximation formula which uses $L_\infty(\mathbb{R}^3)$-norm instead of $L_2(\mathbb{R}^3)$-norm would be desirable. Several other representation formulas for the potential $Q(x)$ given in Saitō [23,24] could be a reasonable starting point for this problem. We can also expect to relax the conditions under which our approximation formulas was proved.

(2) Study of further relations between the S-matrix and the potential. Now that we have established uniqueness in the inverse scattering

problem for the Schrödinger operator with a general short-range potential, we have a one-to-one correspondence between the potentials and the S-matrices. Therefore we have built the foundation in which to ask the following question: given some properties of the S-matrix, determine some properties of $Q(x)$, and vice versa. Friedman [7] studied this problem for the one dimensional Schrödinger operator H_1 given in (1.1) with the boundary condition

$$y(0) \sin \alpha - y'(0) \cos \alpha = 0. \tag{5.1}$$

He used the Gel'fand-Levitan theory to investigate the relation between the potential $q(r)$ and the spectral function

$$F(x,y) = \int_{-\infty}^{\infty} \frac{\sin\sqrt{\lambda}\, x \sin\sqrt{\lambda}\, y}{\lambda}\, d\sigma(\lambda), \tag{5.2}$$

where $\sigma(\lambda)$ is a measure constructed from the spectral measure $\rho(\lambda)$ associated with one dimensional Schrödinger operator (2.1)-(5.1). Since the appearance of his paper [7] in 1957, there does not seem to have been much work along those lines. But we hope that using the results obtained about the function $f(x,k)$ given in (3.1), we shall be able to study the further relations between $f(x,k)$ and the potential $Q(x)$.

(3) The inverse scattering problem for more general potentials. Isozaki-Hitada [9] generalized the Faddeev's result mentioned in §2 to the Schrödinger operator with a general long-range potential. One big problem in the inverse scattering problem for the long-range scattering is that the S-matrix, which depends on the choice of its modifier, cannot be determined uniquely. But the ambiguity in the scattering data is expected to become negligible as the energy number k goes to infinity. Thus it seems that there would be a good chance that our high energy method can be applied with a suitable modification.

Another interesting Schrödinger-type operator would be the Schrödinger operator with a potential which does not necessarily tend to a constant at infinity. Mochizuki-Uchiyama [12-15] developed the spectral and scattering theory for Schrödinger operator with an "oscillatory long-range" potential $Q(x)$. Here $Q(x)$ is assumed to satisfy

$$\begin{cases} Q(x) = O(1) \\ \dfrac{\partial Q}{\partial |x|} = O(|x|^{-1}) \end{cases} \tag{5.3}$$

and

$$\frac{\partial^2 V}{\partial |x|^2} + aV(x) = O(|x|^{-1-\varepsilon}) \tag{5.4}$$

as $|x| \to \infty$ with constants $a \geq 0$ and $\varepsilon > 0$. In addition $Q(x)$ is assumed to behave uniformly as $|x| \to \infty$ (see Mochizuki-Uchiyama [12, §8,(V2-4)]). Saitō [25] studied a potential $Q(x)$ which satisfies only (5.3) and showed that the limiting absorption principle holds for the Schrödinger operator with such a potential $Q(x)$. For such a potential we have a "nonspherical" radiation condition

$$\nabla u - i\sqrt{\lambda}\,\theta u \quad \text{is small at} \quad \infty \tag{5.5}$$

with which the Schrödinger equation

$$\left(-\Delta + Q(x) - \lambda\right)u = f \tag{5.6}$$

has a unique solution for a positive λ. Here

$$\theta = \theta(x,\lambda) = \nabla R \tag{5.7}$$

and $R = R(x,\lambda)$ is a solution of the eikonal equation

$$|\nabla R|^2 = 1 - \frac{Q(x)}{\lambda}\ .$$

Our radiation condition (5.5) is different from the usual one in the sense that θ is not necessarily equal to $\tilde{x} = x/|x|$. The idea and techniques used in the study of this Schrödinger type operator seem to be useful for the study of an acoustic field in an inhomogeneous fluid. Very recently Barles [4] has proved that the above eikonal equation has a smooth solution which has all necessary asymptotic behaviors at infinity under reasonable conditions on $Q(x)$.

ACKNOWLEDGEMENTS

This work is based on the lectures which the author gave in the fall of 1985 at the University of Heidelberg, West Germany, Aarhus University, Denmark and several other universities in West Germany. The author wishes to express his sincere appreciation to all these institutes and especially Professor Willi Jäger of the University of Heidelberg and Professor Erik Balslev of Aarhus University for the invitation and kind hospitality. Also, he wishes to thank the University of Alabama at Birmingham for granting leave of absence during this period.

REFERENCES

[1] Agmon, S.: Spectral properties of Schrödinger operators and scattering theory, Ann. Scuola Nor. Sup. Pisa (4) 2(1975), 151-218.

[2] Agranovich, Z. S. and V. A. Marchenko: The Inverse Problem of Scattering Theory (English translation), Gordon and Breach, New York, 1963.

[3] Amrein, W., J. Jauch and K. Sinha: Scattering theory in quantum mechanics, Lecture Notes and Supplements in Physics, Benjamin, Reading, 1977.

[4] Balres, G. (CEREMADE, Université de Paris IX, Paris, France): Private communications, 1985.

[5] Cheney, M.: Inverse scattering in dimension two, J. Math. Phys. 25(1984), 94-107.

[6] Faddeev, L. D.: Uniqueness of the inverse scattering problem, Vestn. Leningr. Univ. 7(1956), 126-130.

[7] Friedman, A.: On the properties of a singular Sturm-Liouville equation determined by its spectral functions, Michigan Math. J. 4(1957), 137-145.

[8] Gel'fand, I. M. and B. M. Levitan: On the determination of a differential equation by its spectral function, Izv. Akad. Nauk. SSR ser. Math. 15(1951), 309-360 (English translation: Amer. Math. Soc. Transl. ser. 2, 1(1955), 253-304).

[9] Isozaki, H. and H. Kitada: Asymptotic behavior of the scattering amplitude at high energies, to appear in Scientific Papers of the College of Arts and Sciences, The University of Tokyo, 1986.

[10] Kato, T.: Perturbation theory for linear operators, 2nd edition, Springer, New York, 1976.

[11] Kuroda, S. T.: On the existence and the unitary property of the scattering operator, Nuovo Cimento 12(1959), 431-454.

[12] Mochizuki, K. and J. Uchiyama: Radiation conditions and spectral theory for 2-body Schrödinger operators with "oscillating" long-range potentials I, J. Math. Kyoto Univ., 18(1978), 377-408.

[13] _____: Radiation conditions and spectral theory for 2-body Schrödinger operators with "oscillating" long-range potentials II, J. Math. Kyoto Univ., 19(1979), 47-70.

[14] _____: Radiation conditions and spectral theory for 2-body Schrödinger operators with "oscillating" long-range potentials III, J. Math. Kyoto Univ. 21(1981), 605-618.

[15] _____: Time dependent representations of the stationary wave operators for "oscillating" long-range potentials, J. Math. Kyoto Univ., 18(1982), 947-972.

[16] Newton, R.: Inverse scattering, II, J. Math. Phys. 21(1980), 1698-1715.

[17] Newton, R.: Inverse scattering, III, J. Math. Phys. 22(1981), 2191-2200.

[18] _____: Inverse scattering, IV, J. Math. Phys. 23(1982), 592-604.

[19] Prosser, R.: Formal solutions of inverse scattering problems, IV. Error estimates, J. Math. Phys. 23(1982), 2127-2130.

[20] Saitō, Y.: Spectral and scattering theory for second-order differential operators with operator-valued coefficients, Osaka J. Math. 9(1972), 463-498.

[21] _____: On the S-matrix for Schrödinger operators with long-range potentials, J. Reine Angew. Math. 314(1980), 99-116.

[22] _____: Some properties of the scattering amplitude and the inverse scattering problem, Osaka J. Math. 19(1982), 527-547.

[23] _____: An inverse problem in potential theory and the inverse scattering problem, J. Math. Kyoto Univ. 22(1982), 315-329.

[24] _____: An asymptotic behavior of S-matrix and the inverse scattering problem, J. Math. Phys. 25(1984), 3105-3111.

[25] _____: Schrödinger operators with a nonspherical radiation condition, 1985, to appear in Pacific J. Math.

[26] _____: An approximation formula in the inverse scattering problem, 1985, to appear in J. Math. Phys.

α-decay and the exponential law

Erik Skibsted
Matematisk Institut, Universitetsparken,
Ny Munkegade, 8000 Aarhus C,
Denmark

1. Introduction

This paper deals with the problem of the "observability" of the exponential decay law associated with quantum mechanical resonances. It is well-known theoretically that the exponential law cannot hold for all times. Thus the best we can hope for, and what we actually do for sharp resonances (within the framework of a simple model), is to prove the validity of this law for periods of several lifetimes.

Concerning applications we confine ourselves to α-decay from a heavy nucleus. The discussion is based on the following well-known one-dimensional model: An α-particle is considered in a spherically symmetric potential $V = V(r)$ comprising a short-range negative piece from the attraction between nucleons and a positive piece of a longer range from the Coulomb repulsion between protons. The effect of the barrier thus defined by V is that it confines the α-particle to the nucleus for a long period. Within the framework of the above model Gamow [1] and (independently) Gurney and Condon [2] obtained the famous decay rate formula giving lifetime in terms of energy.

According to Gamow the lifetime and energy of the escaping α-particle are identified as follows: For some suitable $k_0 = \alpha - i\beta$ there exists a solution $f(k_0,r)$ of the equation $(-\frac{d^2}{dr^2} + V(r) - k_0^2)$ $f(r) = 0$ on the positive real axis, such that $f(k_0,0) = 0$ and $f(k_0,r) = e^{ik_0 r}$ for r large (k_0 is a resonance in a mathematical sense). $f(k_0,r)$ is assumed to be the α-particle wave function and, putting $E - i\Gamma/2 = k_0^2$, E and Γ^{-1} are the energy and lifetime, respectively. Moreover (more explicitly) the evolution of $f(k_0,r)$ for $t \geq 0$ is given by $e^{-itk_0^2} f(k_0,r)$.

An obvious objection to the theory of Gamow is that $f(k_0,r)$ is not square integrable and thus cannot be a true wave function (possessing energy and lifetime). From a mathematical point of view Gamow's argument, that $f(k_0,r)$ (because Γ/E is small) "looks like" a continuum state (the evolution of which is given by multiplication by e^{-itk^2} for some $k \simeq k_0$), is not acceptable.

In this paper we put Gamow's ideas on a rigorous footing in the following way: For R large (outside the barrier) put $f_R = \chi_{(0,R)} f(k_0,\cdot)$. Then we prove that the evolution of f_R exhibits exponential decay. Furthermore it turns out that the energy and lifetime of f_R are given by E and Γ^{-1}, respectively. More precisely we prove the following (main) result. Let $H = -\frac{d^2}{dr^2} + \ell(\ell+1)/_{r^2} + V(r)$ be the Hamiltonian on $L^2(\mathbb{R}^+)$ (here the potential V is not necessarily compactly supported). Then for R_1 large and $t \geq 0$

$$\| e^{-itH} f_{R_1} - e^{-itk_0^2} f_{R_2}(t) \| \leq K \| f_{R_1} \| \quad , \tag{1,1}$$

where

$$R_2(t) = R_1 + 2\alpha t$$

and

$$K \simeq C\left(\frac{\beta}{\alpha}\right)^{1/4} (t\Gamma)^{1/4} \quad .$$

It is assumed that $\frac{\beta}{\alpha}$ is small.

In the case $\ell = 0$ and assuming that there exists $R_s > 0$ such that $V(r) = 0$ for $r > R_s$, (1.1) holds for all $R_1 > R_s$. Moreover in this case we present two rather different proofs. The statement resulting from the first proof (Section 3) is $C = \pi^{1/2}\, 3^{-3/8}\, 2^{5/4}$, and that from the second proof (Section 4) is $C = 2^{7/4}\, 3^{-1/2}\, \pi^{-1/2}$.

We remark that $R_2(t)$ is the "classical propagation radius" (2α is the classical speed of a free α-particle having the energy E). Furthermore the exponential growth of $f(k_0,r)$ at infinity agrees perfectly with the fact that "the radioactive source is exhausted" (as noted by Gamow too): The interpretation of the increasing probability density is that "larger r corresponds to earlier escape time".

We organize the paper as follows: Section 2 is intended for definitions, and a rather general condition on V (Condition A), which admits a reasonable definition of the concept of resonance, is given. Assuming Condition A we prove in Section 3 the inequality (1.1) for arbitrary $\ell \geq 0$. The proof is given in position space representation. In Section 4 a proof is given in energy space representation (in which H is diagonalized). We assume now that $V(r) = 0$ for $r > R_s$ and that $\ell = 0$. As indicated above the error constant of Section 4 is smaller than that of Section 3. This fact is due to the very explicit representation in energy space representation (expressed in Proposition 4.1), which immediately reduces the proof to a matter of estimating an explicit integral (Lemma 4.5). In Section 5 we present a numerical application to the α-decay problem (based on Corollary 4.8). The exponential decay law is proved to be valid for periods of several lifetimes.

The proof of (1.1) presented in Section 4 can also be found in [4] and [5]. The proof given in Section 3 is based on ideas in [5] and [6]. The paper [6] concerns the general multiplicative case (a result like (1.1) holds also for non-radial potentials). Furthermore a generalization to three-body resonances is given.

2. Definitions and assumptions on V.

Let V be a multiplicative, radial and real potential such that the following condition is satisfied.

Condition A :

1) $\int_0^R r|V(r)|dr < \infty$ for all R positive .

2) There exists $\frac{\pi}{2} > \sigma > 0$, $R_\sigma > 0$ and $a > 0$ such that for $r \geq R_\sigma$ and $\ell \geq 0$

$$\ell(\ell+1)/r^2 + V(r) = V_1(r) + V_2(r) ,$$

where

$$\int_{R_\sigma}^{\infty} V_2(r)\, e^{2ar} dr < \infty$$

and

$V_1(r)$ has a continuous extension to $M_\sigma \equiv \{z \mid |z| \geq R_\sigma$ and $|\text{Arg}\, z| \leq \sigma\}$, analytic in the interior of M_σ ; furthermore, $z V_1(z) \to 0$ for $z \to \infty$ in M_σ and

$$\sup_{-\sigma \leq \theta \leq \sigma} \int_{e^{i\theta}R_\sigma}^{e^{i\theta}\infty} |dz||V_1(z)| < \infty .$$

Let for all $\ell \geq 0$ $H_0^\ell = -\frac{d^2}{dr^2} + \ell(\ell+1)/r^2$ denote the free Hamiltonian on $L^2(\mathbb{R}^+)$ (determined in the case $\ell = 0$ by the boundary condition $g(0) = 0$ for $g \in D(H_0^\ell)$). Then it is easy to prove that V is infinitesimally form-bounded with respect to H_0^ℓ. Hence we can construct the total Hamiltonian $H^\ell = H_0^\ell + V$ by the standard quadratic form technique.

The following solutions $\varphi_\ell(k,r)$, $f_\ell(k,r)$, $\Psi_\ell^+(k,r)$ of the equation on $\mathbb{R}^+$ $\left(-\frac{d^2}{dr^2} + \ell(\ell+1)/r^2 + V(r) - k^2\right)\Psi(r) = 0$ are all discussed in Newton [3] Sections 12.1 - 2. $\varphi_\ell(k,r)$, the regular solution defined for all k , satisfies $\lim_{r\to 0} r^{-(\ell+1)} \varphi_\ell(k,r) = 1$. $f_\ell(k,r)$ is for $k \in \mathbb{C}^+ \equiv \{\zeta \neq 0 \mid \text{Im}\,\zeta \geq 0\}$ the outgoing solution defined uniquely for all $r > 0$ by

$$f_\ell(k,r) = e^{ikr} - \frac{1}{k}\int_r^\infty dr' \sin k(r-r')\{\ell(\ell+1)/r'^2 + V(r')\} f_\ell(k,r') .$$

For $k \in \mathbb{C}^+$ the Jost function $F_\ell(k)$ is given by $F_\ell(k) = W(f_\ell(k,r), \varphi_\ell(k,r))$, where W denotes the Wronskian. For $k \in \mathbb{R}^+$ the physical wave function $\Psi_\ell^+(k,r)$ is defined to be equal to $\frac{k}{F_\ell(k)} \varphi_\ell(k,r) e^{i\ell\pi/2}$. It is known that $F_\ell(k) \neq 0$ $(k \in \mathbb{R}^+)$.

In Section 4 we shall make use of $\Psi_\ell^+(k,r)$ and (2,1) - (2.4) given below.

For $k \in \mathbb{R}\setminus\{0\}$ we have

$$\varphi_\ell(k,r) = \frac{1}{2ik}(F_\ell(-k) f_\ell(k,r) - F_\ell(k) f_\ell(-k,r)) . \tag{2.1}$$

Put $S_\ell(k) = (-1)^\ell \frac{F_\ell(-k)}{F_\ell(k)}$ $(k \in \mathbb{R}\setminus\{0\})$.

Then for all $k \in \mathbb{R}\setminus\{0\}$ (because $\overline{F_\ell(k)} = F_\ell(-\bar{k})$),

$$|S_\ell(k)| = 1 \tag{2.2}$$

and

$$\overline{\Psi_\ell^+(k,r)} = \frac{k\,\varphi_\ell(k,r)}{F_\ell(-k)} e^{-i\ell\pi/2} . \tag{2.3}$$

The kernel of the spectral density of $\frac{d}{d\lambda}E_\lambda^\ell$ of H^ℓ is given by

$$\frac{d}{d\lambda} E_\lambda^\ell(r,r') = \frac{1}{k\pi} \Psi_\ell^+(k,r) \overline{\Psi_\ell^+(k,r')} , \tag{2.4}$$

where we have put $\lambda = k^2$ for $k > 0$.

See [4]. A non-rigorous discussion is given in [3] 12.1.5,

The orthogonal projections onto the subspace of absolute continuity of H^ℓ and the span of all eigenvectors of H^ℓ are given by $P_{ac}^\ell = I - E_0^\ell$ and $P_e^\ell = E_0^\ell = \chi_{(-\infty,0]}(H^\ell)$, respectively.

It is well-known that $\varphi_\ell(k,r)$ is entire analytic in k for all fixed $r > 0$. Concerning $f_\ell(k,r)$ and $F_\ell(k)$, it is proved in [5] that these functions have analytic continuations from $\mathbb{C}^+$ to

$\mathbb{C}^+ \cup (S_\sigma \cap T_a)$, where $S_\sigma \equiv \{\zeta \neq 0 \mid 0 > \operatorname{Arg}\zeta > -\sigma\}$ and $T_a \equiv \{\zeta \mid 0 > \operatorname{Im}\zeta > -a\}$. (2.1) holds true for all $k \in S_\sigma \cap T_a$ and $r > 0$.

A resonance: If $F_\ell(k_0) = 0$ for a point $k_0 = \alpha - i\beta \in S_\sigma \cap T_a$, we call k_0 a resonance.

The Gamow function: Given a resonance k_0 , we call $f_\ell(k_0,r)$ the Gamow function.

We define the resonance energy E and width Γ by $k_0^2 = \alpha^2 - \beta^2 - i2\alpha\beta = E - i\Gamma/_2$.

The following Wronski formulas are useful (remark that according to (2.1), $\frac{d}{dr} f_\ell(k_0,r)$ is absolutely continuous on $\mathbb{R}^+$, and a.e. $(-\frac{d^2}{dr^2} + V(r) - k_0^2) f_\ell(k_0,r) = 0$).

$$\frac{d}{dr} W(\overline{f_\ell(k_0,r)}, f_\ell(k_0,r)) = (\overline{k_0^2} - k_0^2) \; |f_\ell(k_0,r)|^2 \; . \qquad (2.5)$$

$$\frac{d}{dr} W(\varphi_\ell(k,r) \, , f_\ell(k_0,r)) = (k^2 - k_0^2) \, \varphi_\ell(k,r) f_\ell(k_0,r) \; . \; (2.6)$$

Let $S_{[0,1]} = \{\varphi \in C^1([0,1]) \mid \varphi'' \in L^2((0,1)) , \; \varphi(1) = \varphi'(1) = \varphi'(0) = 0$ and $\varphi(0) = 1\}$. For $\varphi \in S_{[0,1]}$ we write $\|\varphi\|$, $\|\varphi'\|$ and $\|\varphi''\|$ instead of $\|\varphi\|_{L^2}$, $\|\varphi'\|_{L^2}$ and $\|\varphi''\|_{L^2}$, respectively.

Let for all fixed $d,R > 0$ and $\varphi \in S_{[0,1]}$ the function $\varphi_R : \mathbb{R}^+ \to \mathbb{C}$ be given by

$$\varphi_R(r) = \begin{cases} 1 & , \text{ for } r < R \\ \varphi\left(\frac{r-R}{d}\right) & , \text{ for } R \le r \le R+d \\ 0 & , \text{ for } R+d < r \; . \end{cases}$$

Given $d,R > 0$, $\varphi \in S_{[0,1]}$ and a Gamow function we define functions f_R , g_R , $\Psi_R \in L^2(\mathbb{R}^+)$ as follows:

$$f_R = f_\ell(k_0,\cdot)\chi_{(0,R)} \quad ,$$

$$g_R = f_\ell(k_0,\cdot)\ \varphi_R$$

and

$$\Psi_R = -\frac{d}{dr^2}\ \varphi_R f_\ell(k_0,\cdot) - 2\frac{d}{dr}\ \varphi_R \frac{d}{dr} f_\ell(k_0,\cdot) \ .$$

It is easily verified that $g_R \in D(H^\ell)$ and that $(H^\ell - k_0^2)g_R = \Psi_R$.

Given a resonance we define functions $\varepsilon_1, \varepsilon_2 : \mathbb{R}^+ \to \mathbb{C}$ by the equations

$$f_\ell(k_0,r) = e^{ik_0r}(1 + \varepsilon_1(r))$$

and

$$\frac{d}{dr} f_\ell(k_0,r) = i\,k_0\, e^{ik_0r}(1 + \varepsilon_2(r)) \ .$$

An error function: We put $\varepsilon(r) = \sup_{r' \geq r} \max\{|\varepsilon_1(r')|\ ,\ |\varepsilon_2(r')|\}$.

For $r > R_\sigma$ it is proved in [5] that

$$\varepsilon(r) \leq B_{V_1}(r,k_0) - 1 + B_{V_1}(r,k_0)(\exp\{\frac{1}{k_0}(B_{V_1}(r,k_0))^2 C_{V_2}(r)\} - 1) \ ,$$

where, putting $\gamma(r) = r(1 + i[0,\mathrm{tg}\sigma]) \cup r(1 + i\,\mathrm{tg}\sigma)(1,\infty)$ considered as a path from r to ∞ ,

$$B_{V_1}(r,k_0) = \exp\{\frac{1}{|k_0|}\max\left(\sup_{r' \geq r}\int_{\gamma(r')}|dz||V_1(z)|,\ \int_r^\infty dr'|V_1(r')|\right)$$

and

$$C_{V_2}(r) = \int_r^\infty dr'\ e^{2ar'}|V_2(r')| \ .$$

In particular $\varepsilon(r) \to 0$ for $r \to \infty$.

Let R_0 be an arbitrary positive number, such that $\frac{|k_0|}{\alpha}(2\varepsilon(R_0) + (\varepsilon(R_0))^2) < 1$.

3. A proof in position space representation.

In this Section Condition A is assumed. We consider a fixed ℓ - wave resonance $k_0 = \alpha - i\beta$.

Proposition 3.1. For all $R > 0$

$$\frac{e^{2\beta R}}{2\beta}\{1 - \frac{|k_0|}{\alpha}(2\varepsilon(R) + (\varepsilon(R))^2)\} \leq \|f_R\|^2 \leq \frac{e^{2\beta R}}{2\beta}\{1 + \frac{|k_0|}{\alpha}(2\varepsilon(R) + (\varepsilon(R))^2)\}$$

Proof. We integrate (2.5) and obtain (because $f_\ell(k_0,0) = 0$),

$$i\Gamma\|f_R\|^2 = 2i\,\mathrm{Im}\{\frac{d}{dr}f_\ell(k_0,R)\overline{f_\ell(k_0,R)}\}$$
$$= 2i\,\mathrm{Im}\{ik_0 e^{2\beta R}(1 + \varepsilon_2(R))(1 + \overline{\varepsilon_1(R)})\}.$$

Hence

$$\|f_R\|^2 = \frac{e^{2\beta R}}{2\beta}(1 + \mathrm{Re}\{\frac{|k_0|}{\alpha}(\varepsilon_2(R) + \overline{\varepsilon_1(R)} + \varepsilon_2(R)\overline{\varepsilon_1(R)})\}),$$

and the Lemma follows.

Lemma 3.2. Let $d > 0$, $\varphi \in S_{[0,1]}$ and $R_1 > R_0$ be given. Then, putting $R_2(s) = 2\alpha s + R_1$, we have that the $L^2(\mathbb{R}^+)$-valued function $e^{is(H^\ell - k_0^2)}g_{R_2(s)}$ is continuously differentiable for all $s \geq 0$. Moreover

$$\|\frac{d}{ds}\{e^{is(H^\ell - k_0^2)}g_{R_2(s)}\}\| \leq \|f_{R_1}\|\,C_1 ,$$

where

$$C_1 = \{1 - \frac{|k_0|}{\alpha}(2\varepsilon(R_1) + (\varepsilon(R_1))^2)\}^{-\frac{1}{2}}e^{\beta d}(2\beta d)^{\frac{1}{2}}$$

$$\{\|\varphi''/d^2 + \frac{2\beta}{d}\varphi'\| + \varepsilon(R_2(s))(\|\varphi''/d^2 - i\frac{2\alpha}{d}\varphi'\| + 2\frac{|k_0|}{d}\|\varphi'\|)\} .$$

Proof. For all $s \geq 0$ and $r \in \mathbb{R}^+$

$$\frac{d}{ds}\varphi_{R_2(s)}(r) = \chi_{(R_2(s),R_2(s)+d)}(r)\,\frac{(-2\alpha)}{d}\,\varphi'\left(\frac{r - R_2(s)}{d}\right) . \qquad (3.1)$$

Using (3.1) one verifies that the derivative of $e^{is(H^{\ell}-k_0^2)} g_{R_2(s)}$ exists and is given by

$$\frac{d}{ds}\{e^{is(H^{\ell}-k_0^2)} g_{R_2(s)}\} = e^{is(H^{\ell}-k_0^2)} \{i\,\Psi_{R_2(s)} + \frac{d}{ds}\varphi_{R_2(s)} f_{\ell}(k_0,\cdot)\} . \quad (3.2)$$

Clearly a.e. $i\,\Psi_{R_2(s)}(r) + \frac{d}{ds}\varphi_{R_2(s)}(r) f_{\ell}(k_0,r) = A+B+C+D+E+F$,

where, putting $R_2 = R_2(s)$ and $\chi(r) = \chi_{(R_2,R_2+d)}(r)$,

$$A = -i\,\frac{1}{d^2}\,\varphi''\left(\frac{r-R_2}{d}\right)\chi(r)e^{ik_0 r} ,$$

$$B = -i\,\frac{1}{d^2}\,\varphi''\left(\frac{r-R_2}{d}\right)\chi(r)e^{ik_0 r}\,\varepsilon_1(r) ,$$

$$C = -i\,\frac{2}{d}\,\varphi'\left(\frac{r-R_2}{d}\right)\chi(r)\,ik_0 e^{ik_0 r} ,$$

$$D = -i\,\frac{2}{d}\,\varphi'\left(\frac{r-R_2}{d}\right)\chi(r)\,ik_0 e^{ik_0 r}\,\varepsilon_2(r) ,$$

$$E = -\,\frac{2\alpha}{d}\,\varphi'\left(\frac{r-R_2}{d}\right)\chi(r)e^{ik_0 r} ,$$

and

$$F = -\,\frac{2\alpha}{d}\,\varphi'\left(\frac{r-R_2}{d}\right)\chi(r)e^{ik_0 r}\,\varepsilon_1(r) .$$

We remark that C and E "almost cancel" (a crucial point in our proof !):

$$C + E = -i\,\frac{2\beta}{d}\,\varphi'\left(\frac{r-R_2}{d}\right)\chi(r)e^{ik_0 r} .$$

Obviously the first part of the Lemma follows from the above expressions and (3.2).

Now we use the estimates

$$\|A+C+E\| \le e^{\beta(R_2+d)}\sqrt{d}\,\|\varphi''/d^2 + \frac{2\beta}{d}\varphi'\| ,$$

$$\|B+F\| \le e^{\beta(R_2+d)}\sqrt{d}\,\|\varphi''/d^2 - i\frac{2\alpha}{d}\varphi'\|\,\varepsilon(R_2) ,$$

$$\|D\| \le \frac{2|k_0|}{d}\,e^{\beta(R_2+d)}\sqrt{d}\,\|\varphi'\|\,\varepsilon(R_2) ,$$

and obtain from (3.2) and Proposition 3.1, that

$$\| \frac{d}{ds} \{e^{is(H^{\ell}-k_0^2)} g_{R_2(s)}\| \leq \frac{e^{\beta R_1}}{\sqrt{2\beta}} e^{\beta d}(2\beta d)^{\frac{1}{2}}$$

$$\{\|\varphi''/d^2 + \frac{2\beta}{d}\varphi'\| + \varepsilon(R_2)(\|\varphi''/d^2 - i\frac{2\alpha}{d}\varphi'\| + \frac{2|k_0|}{d}\|\varphi'\|)\}$$

$$\leq \|f_{R_1}\|\{1 - \frac{|k_0|}{\alpha}(2\varepsilon(R_1) + (\varepsilon(R_1))^2)\}^{-\frac{1}{2}} e^{\beta d}(2\beta d)^{\frac{1}{2}}$$

$$\{\|\varphi''/d^2 + \frac{2\beta}{d}\varphi'\| + \varepsilon(R_2)(\|\varphi''/d^2 - i\frac{2\alpha}{d}\varphi'\| + \frac{2|k_0|}{d}\|\varphi'\|)\} .$$

The Lemma is proved.

Our main result is the following.

<u>Theorem 3.3</u>. For all $t \geq 0$ and $R_1 > R_0$

$$\|e^{-itH^{\ell}} f_{R_1} - e^{-itk_0^2} f_{R_2(t)}\| \leq \|f_{R_1}\| K_1 ,$$

where

$$R_2(t) = R_1 + 2\alpha t$$

and

$$K_1 = K_1(t,R_1,k_0) = \pi^{\frac{1}{2}} 3^{-3/8} 2^{5/4} \left(\frac{\beta}{\alpha}\right)^{1/4} (t\Gamma)^{1/4}$$

$$\exp\{\pi 3^{1/4} 8^{-\frac{1}{2}}\left(\frac{\beta}{\alpha}\right)^{\frac{1}{2}}(t\Gamma)^{\frac{1}{2}}\}\{1 - \frac{|k_0|}{\alpha}(2\varepsilon(R_1) + (\varepsilon(R_1))^2)\}^{-\frac{1}{2}}$$

$$\{1 + \varepsilon(R_1) + \left(\frac{\beta}{\alpha}\right)^{\frac{1}{2}}(t\Gamma)^{\frac{1}{2}} 3^{1/4} 2^{-5/2} + 3^{1/4} 2^{-\frac{1}{2}}|k_0| t^{-\frac{1}{2}} \int_0^t ds\,\varepsilon(R_2(s))\} .$$

<u>Proof</u>. Let $t > 0$ and $R_1 > R_0$ be given, and fix $d > 0$ and $\varphi \in S_{[0,1]}$ arbitrarily.

For all $r, R > 0$ we have that

$$f_R(r) - g_R(r) = -\chi_{(R,R+d)}(r)\varphi\left(\frac{r-R}{d}\right)e^{ik_0 r}(1 + \varepsilon_1(r)) .$$

Hence for all $R > 0$

$$\|f_R - g_R\| \leq e^{\beta(R+d)}(1 + \varepsilon(R))\sqrt{d}\,\|\varphi\| ,$$

and thus in particular, by Proposition 3.1,

$$\|f_{R_1}-g_{R_1}\| \le \|f_{R_1}\|\{1-\frac{|k_0|}{\alpha}(2\varepsilon(R_1)+(\varepsilon(R_1))^2)\}^{-\frac{1}{2}}e^{\beta d}(2\beta d)^{\frac{1}{2}}(1+\varepsilon(R_1))\|\varphi\|, \quad (3.3)$$

and

$$\|f_{R_2(t)}-g_{R_2(t)}\| \le \|f_{R_1}\|\{1-\frac{|k_0|}{\alpha}(2\varepsilon(R_1)+(\varepsilon(R_1))^2)\}^{-\frac{1}{2}}e^{\Gamma t/2}e^{\beta d}(2\beta d)^{\frac{1}{2}}(1+\varepsilon(R_1))\|\varphi\|. \quad (3.4)$$

By the first part of Lemma 3.2

$$\|e^{-itH^{\ell}}f_{R_1}-e^{-itk_0^2}f_{R_2(t)}\| \le \|e^{-itH^{\ell}}(f_{R_1}-g_{R_1})\|$$

$$+\ \|g_{R_1}-e^{it(H^{\ell}-k_0^2)}g_{R_2(t)}\| + \|e^{-itk_0^2}(g_{R_2(t)}-f_{R_2(t)})\|$$

$$\le\ \|f_{R_1}-g_{R_1}\| + \int_0^t ds\,\|\frac{d}{ds}\{e^{is(H^{\ell}-k_0^2)}g_{R_2(s)}\}\| + e^{-\Gamma t/2}\|f_{R_2(t)}-g_{R_2(t)}\| .$$

Now we apply (3.3), (3.4) and the second part of Lemma 3.2 and obtain

$$\|e^{-itH^{\ell}}f_{R_1}-e^{-itk_0^2}f_{R_2(t)}\| \le \|f_{R_1}\|\{1-\frac{|k_0|}{\alpha}(2\varepsilon(R_1)+(\varepsilon(R_1))^2)\}^{-\frac{1}{2}}$$

$$e^{\beta d}(2\beta d)^{\frac{1}{2}}\{2(1+\varepsilon(R_1))\|\varphi\| + t\|\varphi''/d^2+\frac{2\beta}{d}\varphi'\| + \quad (3.5)$$

$$(\|\varphi''/d^2-\frac{i2\alpha}{d}\varphi'\| + \frac{2|k_0|}{d}\|\varphi'\|)\int_0^t ds\,\varepsilon(R_2(s))\} .$$

It follows from (3.5) that

$$\|e^{-itH^{\ell}}f_{R_1}-e^{-itk_0^2}f_{R_2(t)}\| \le \|f_{R_1}\|\{1-\frac{|k_0|}{\alpha}(2\varepsilon(R_1)+(\varepsilon(R_1))^2)\}^{-\frac{1}{2}}$$

$$e^{\beta d}(2\beta d)^{\frac{1}{2}}\{2(1+\varepsilon(R_1))\|\varphi\| + t/_{d^2}\|\varphi''\|(1+2\beta d\frac{\|\varphi'\|}{\|\varphi''\|}) +$$

$$t/_{d^2}\|\varphi''\|\varepsilon(R_1) + \frac{4|k_0|}{d}\|\varphi'\|\int_0^t ds\,\varepsilon(R_2(s))\} .$$

The inequality holds true for all $d>0$ and $\varphi \in S_{[0,1]}$. Hence in particular taking $d = \left(\frac{3}{2}\right)^{\frac{1}{2}} t^{\frac{1}{2}} \|\varphi''\|^{\frac{1}{2}}\|\varphi\|^{-\frac{1}{2}}$

$$\| e^{-itH^{\ell}} f_{R_1} - e^{-itk_0^2} f_{R_2(t)} \| \leq \| f_{R_1} \| \{ 1 - \frac{|k_0|}{d} (2\varepsilon(R_1) + (\varepsilon(R_1))^2) \}^{-\frac{1}{2}}$$

$$\exp\{ \left(\frac{\beta}{\alpha}\right)^{\frac{1}{2}} (t\Gamma)^{\frac{1}{2}} \left(\frac{3}{2}\right)^{\frac{1}{2}} 2^{-1} \| \varphi'' \|^{\frac{1}{2}} \| \varphi \|^{-\frac{1}{2}} \} \left(\frac{3}{2}\right)^{1/4} \left(\frac{\beta}{\alpha}\right)^{1/4} (t\Gamma)^{1/4} \| \varphi'' \|^{1/4} \| \varphi \|^{3/4}$$

$$\{ 2(1 + \varepsilon(R_1)) + \frac{2}{3} (1 + \left(\frac{3}{2}\right)^{\frac{1}{2}} \left(\frac{\beta}{\alpha}\right)^{\frac{1}{2}} (t\Gamma)^{\frac{1}{2}} \| \varphi' \| \, \| \varphi'' \|^{-\frac{1}{2}} \| \varphi \|^{-\frac{1}{2}}) + \frac{2}{3} \varepsilon(R_1) +$$

$$+ 4 \left(\frac{2}{3}\right)^{\frac{1}{2}} |k_0| t^{-\frac{1}{2}} \| \varphi' \| \, \| \varphi'' \|^{-\frac{1}{2}} \| \varphi \|^{-\frac{1}{2}} \int_0^t ds\, \varepsilon(R_2(s)) \} . \tag{3.6}$$

Finally we insert $\varphi \in S_{[0,1]}$, such that $\| \varphi'' \|^{1/4} \| \varphi \|^{3/4}$ is "small" (cf. Remark 3.4 below), into the right-hand side of (3.6). We choose φ given by $\varphi(x) = \cos^2(\frac{x\pi}{2})$, and find in this case that $\| \varphi \|$, $\| \varphi' \|$ and $\| \varphi'' \|$ are equal to $\left(\frac{3}{8}\right)^{\frac{1}{2}}$, $\pi 8^{-\frac{1}{2}}$ and $\pi^2 8^{-\frac{1}{2}}$, respectively.

The proof is complete.

<u>Remark 3.4</u>. Consider the functional $[\cdot]$ on $S_{[0,1]}$ given by $[\varphi] = \| \varphi'' \|^{1/4} \| \varphi \|^{3/4}$. In the proof of Theorem 3.3 appears the problem to minimize $[\cdot]$. It can be proved that $[\varphi] > 2^{-\frac{1}{2}}$ for all $\varphi \in S_{[0,1]}$. However we are not able to determine the greatest lower bound. Better choices of φ than $\varphi_4(x) = \cos^2(\frac{x\pi}{2})$ are given by $\varphi_n(x) = \cos^{\frac{n}{2}}(\frac{x\pi}{2})$ for $n = 5,6,\ldots$. In fact

$$[\varphi_4] > [\varphi_5] > \ldots > \lim_{n\to\infty} [\varphi_n] = \pi^{1/4} 3^{1/8} 8^{-1/4} .$$

<u>Corollary 3.5</u>. For all $t \geq 0$ and $R_1 > R_0$

$$(e^{-itH^{\ell}} f_{R_1}, \chi_{(0,R_1)} e^{-itH^{\ell}} f_{R_1}) \| f_{R_1} \|^{-2} = e^{-\Gamma t}(1 + y_1) ,$$

where $|y_1| \leq e^{\Gamma t} K_1^2 + 2e^{\Gamma t/2} K_1$.

Proof. We use that

$$\| \chi_{(0,R_1)} e^{-itH^\ell} f_{R_1} \|^2 = e^{-\Gamma t} \| \chi_{(0,R_1)} f_{R_2(t)} \|^2 +$$

$$\| \chi_{(0,R_1)} \{ e^{-itH^\ell} f_{R_1} - e^{-itk_0^2} f_{R_2(t)} \} \|^2 +$$

$$+ 2\,\mathrm{Re}(\{ e^{-itH^\ell} f_{R_1} - e^{-itk_0^2} f_{R_2(t)} \}, \chi_{(0,R_1)} e^{-itk_0^2} f_{R_2(t)}).$$

The following theorem gives information on the energy localization of the state f_{R_1}. A more explicit result appears in Section 4 (cf. Remark 4.4).

Theorem 3.6. For all $n > 0$ and $R_1 > R_0$

$$\| \int_{I_n} dE_\lambda^\ell f_{R_1} \|^2 \leq \| f_{R_1} \|^2 K_2$$

where

$$I_n = \{ \lambda \in \mathbb{R} \mid |\lambda - E| \geq n \frac{\Gamma}{2} \}$$

and

$$K_2 = K_2(n, R_1, k_0) = \pi 3^{\frac{1}{2}} \frac{|k_0|}{\alpha} (n^2+1)^{-\frac{1}{2}} \exp\{ 2\pi 3^{-\frac{1}{2}} \frac{|k_0|}{\alpha} (n^2+1)^{-\frac{1}{2}} \}$$

$$\{ 1 + 3^{\frac{1}{2}} 4^{-1} \frac{\beta}{\alpha} (n^2+1)^{\frac{1}{2}} \}^2 (1 + \varepsilon(R_1))^2 \{ 1 - \frac{|k_0|}{\alpha} (2\varepsilon(R_1) + (\varepsilon(R_1))^2) \}^{-1} .$$

Proof. Let $n > 0$ and $R_1 > R_0$ be given, and fix $d > 0$ and $\varphi \in S_{[0,1]}$ arbitrarily.

The identity

$$g_{R_1} = (H^\ell - k_0^2)^{-1} \Psi_{R_1}$$

and an application of the spectral theorem provide that for all real x,

$$\int_{(-\infty,x]} dE_\lambda^\ell g_{R_1} = \int_{(-\infty,x]} (\lambda - k_0^2)^{-1} d\, E_\lambda^\ell \Psi_{R_1} . \tag{3.7}$$

From (3.7) we obtain, applying the spectral theorem again, the following inequality

$$\| \int_{I_n} d E_\lambda^\ell f_{R_1} \| \leq \| f_{R_1} - g_{R_1} \| + \frac{2}{\Gamma}(n^2+1)^{-\frac{1}{2}} \| \Psi_{R_1} \| \; . \tag{3.8}$$

For a.e. $r \in \mathbb{R}^+$

$$\Psi_{R_1}(r) = - \chi_{(R_1, R_1+d)}(r) e^{ik_0 r} \left\{ \frac{1}{d^2} \varphi'' \left(\frac{r - R_1}{d} \right) (1 + \varepsilon_1(r)) + \frac{2}{d} \varphi' \left(\frac{r - R_1}{d} \right) i k_0 (1 + \varepsilon_2(r)) \right\} .$$

Hence using Proposition 3.1,

$$\| \Psi_{R_1} \| \leq e^{\beta(R_1+d)} \sqrt{d} \, \{ \| \varphi''/_{d^2} + \frac{2ik_0}{d} \varphi' \| + \varepsilon(R_1) (\| \varphi'' \| /_{d^2} + \frac{2|k_0|}{d} \| \varphi' \|) \}$$

$$\leq \| f_{R_1} \| \{ 1 - \frac{|k_0|}{\alpha} (2\varepsilon(R_1) + (\varepsilon(R_1))^2) \}^{-\frac{1}{2}} e^{\beta d} (2\beta d)^{\frac{1}{2}} \tag{3.9}$$

$$(\| \varphi'' \| /_{d^2} + \frac{2|k_0|}{d} \| \varphi' \|) (1 + \varepsilon(R_1)) .$$

We now apply (3.3) and (3.9) to the right-hand side of (3.8) and obtain

$$\| \int_{I_n} d E_\lambda^\ell f_{R_1} \| \leq \| f_{R_1} \| e^{\beta d} (2\beta d)^{\frac{1}{2}} \{ 1 - \frac{|k_0|}{\alpha} (2\varepsilon(R_1) + (\varepsilon(R_1))^2) \}^{-\frac{1}{2}}$$
$$(1 + \varepsilon(R_1)) \{ \| \varphi \| + (\| \varphi'' \| /_{d^2} + \frac{2|k_0|}{d} \| \varphi' \|) \frac{2}{\Gamma} (n^2+1)^{-\frac{1}{2}} \} . \tag{3.10}$$

Finally we note that (3.10) holds true for $d = 4|k_0| \, \Gamma^{-1} (n^2+1)^{-\frac{1}{2}} \| \varphi' \| \, \| \varphi \|^{-1}$. Choosing $\varphi(x) = \cos^2 (\frac{x\pi}{2})$ (as in the proof of Theorem 3.3), we easily complete the proof.

<u>Remark 3.7</u>. In the proof of Theorem 3.6 one would like to minimize (cf. Remark 3.4) the functional on $S_{[0,1]}$ given by $[\varphi] = \| \varphi \| \, \| \varphi' \|$. Because $-1 = 2 \, \mathrm{Re}(\varphi, \varphi')$, $[\varphi] > \frac{1}{2}$ for all $\varphi \in S_{[0,1]}$.

4. A proof in energy space representation.

In this Section Condition A is replaced by the following stronger condition: For some $R_s > 0$, $V(r) = 0$ for $r > R_s$ and $\int_0^{R_s} r\,|V(r)|\,dr < \infty$.

We consider a fixed s-wave resonance k_0 .

The proofs can be generalized to the case of Condition A and ℓ-wave resonances defined for arbitrary ℓ [5].

For convenience we put $H = H^0$, $\Psi^+(k,r) = \Psi_0^+(k,r)$, etc.

Proposition 4.1. For all $k > 0$ and $R > R_s$

$$< \Psi^+(k,\cdot) , f_R > \equiv \int_0^\infty \overline{\Psi^+(k,r)}\, f_R(r)\,dr =$$

$$\frac{-k}{k^2 - k_0^2} S(-k) e^{i(k_0-k)R} [1 + \frac{k - k_0}{2k} (S(k) e^{2ikR} - 1)] .$$

Proof. We integrate (2.6) and obtain (because $f(k_0,0) = 0$)

$$\int_0^R \varphi(k,r) f(k_0,r)\,dr = \frac{-1}{k^2 - k_0^2} (-\varphi(k,R) \frac{d}{dr} f(k_0,R) + \frac{d}{dr} \varphi(k,R) f(k_0,R)) . \quad (4.1)$$

The Proposition follows easily from (4.1), (2.1) and (2.3) (note that $f(k,r) = e^{ikr}$ for $r > R_s$ and $\pm k > 0$ or $k = k_0$).

Lemma 4.2. For all $R > R_s$

$$\| P_e f_R \|^2 \leq \| f_R \|^2 C_2 ,$$

where

$$C_2 = \pi\, 3^{\frac{1}{2}}\, 2\{1 + 3^{\frac{1}{2}}\, 8^{-1} (1 + \left(\frac{\beta}{\alpha}\right)^2)\}^2 \frac{\beta}{\alpha} \exp\{4\pi\, 3^{-\frac{1}{2}} \frac{\beta}{\alpha}\} .$$

Proof. Put $n = 2 E \Gamma^{-1}$. Then $\| P_e f_R \|^2 \leq \| \int_{I_n} d E_\lambda f_R \|^2$, and the Lemma follows from Theorem 3.6 (note that $(n^2 + 1)^{\frac{1}{2}} = \frac{1}{2} \frac{\alpha}{\beta} (1 + \left(\frac{\beta}{\alpha}\right)^2)$ and hence that $\frac{|k_0|}{\alpha} (n^2 + 1)^{-\frac{1}{2}} < 2 \frac{\beta}{\alpha}$) .

Remark 4.3. A similar (weaker) result is proved in [4] using the Propositions 3.1 and 4.1 together with (2.4).

Remark 4.4. Lemma 4.2 states that f_R for $R > R_s$ is essentially a continuum state. Hence, by Proposition 4.1, the energy distribution of f_R is given $(\lambda = k^2)$ by

$$|<\Psi^+(\lambda^{\frac{1}{2}},\cdot), f_R>|^2 \frac{1}{\pi}\lambda^{-\frac{1}{2}} \simeq \frac{1}{\pi} \frac{\lambda^{\frac{1}{2}}}{(\lambda - E)^2 + \left(\frac{\Gamma}{2}\right)^2} e^{2\beta R}$$

(a Breit-Wigner form). We have used (2.2). This result is more explicit than Theorem 3.6. Moreover a new proof of this Theorem can be given, and one can verify that the form of the constant K_2 in Theorem 3.6 $(K_2 \simeq C n^{-1})$ is "optimal". (On the other hand $C = \pi 3^{\frac{1}{2}}$ is too large; $C = \frac{2}{\pi}$ is "optimal").

Lemma 4.5. Let $R_1 > R_s$ and $t \geq 0$ be given and put R_2 $R_2(t) = 2\alpha t + R_1$. Then

$$\| P_{ac}(e^{-itH} f_{R_1} - e^{-itk_0^2} f_{R_2(t)})\|^2 \leq \| f_{R_1}\|^2 C_3 ,$$

where

$$C_3 = \frac{16}{\pi} \{ \left(\frac{\beta}{\alpha}\right)^{1/4} \{2^{-\frac{1}{2}} 3^{-1} (t\Gamma)^{\frac{1}{2}} (1 + \left(\frac{\beta}{\alpha}\right)^{1/4})^2 (1 + \frac{3}{32} t\Gamma \frac{\beta}{\alpha}) + \left(\frac{\beta}{\alpha}\right)^{1/4}\}^{\frac{1}{2}} + (\frac{\beta}{\alpha})^{\frac{1}{2}}\}^2 .$$

Proof. Put $R_2 = R_2(t)$. According to Proposition 4.1 we have for all $k > 0$ that

$$<\Psi^+(k,\cdot), e^{-itk^2} f_{R_1}> - <\Psi^+(k,\cdot), e^{-itk_0^2} f_{R_2}> = -\frac{k}{k^2 - k_0^2} S(-k)\{a+b+c\} ,$$

where

$$a = e^{-itk^2} e^{i(k_0-k)R_1} \frac{k - k_0}{2k} (S(k) e^{2ikR_1} - 1) ,$$

$$b = -e^{-itk_0^2} e^{i(k_0-k)R_2} \frac{k - k_0}{2k} (S(k) e^{2ikR_2} - 1)$$

and

$$c = e^{-itk^2} e^{i(k_0-k)R_1} (1 - e^{it(k^2 - k_0^2)} e^{i(k_0-k)2\alpha t}) .$$

Using (2.2) and (2.4) we find that

$$\| P_{ac}(e^{-itH} f_{R_1} - e^{-itk_0^2} f_{R_2}) \|^2 \tag{4.2}$$

$$\leq \frac{2}{\pi}\left(\left\{\int_0^\infty dk \frac{k^2}{|k^2-k_0^2|^2}|a|^2\right\}^{\frac{1}{2}} + \left\{\int_0^\infty dk \frac{k^2}{|k^2-k_0^2|^2}|b|^2\right\}^{\frac{1}{2}} + \left\{\int_0^\infty dk \frac{k^2}{|k^2-k_0^2|^2}|c|^2\right\}^{\frac{1}{2}}\right)^2 .$$

Because $|a|^2, |b|^2 \leq e^{2\beta R_1}\left|\frac{k-k_0}{k}\right|^2$, we have the estimates

$$\int_0^\infty dk \frac{k^2}{|k^2-k_0^2|^2}|a|^2 \leq e^{2\beta R_1}\alpha^{-1} \tag{4.3}$$

and

$$\int_0^\infty dk \frac{k^2}{|k^2-k_0^2|^2}|b|^2 \leq e^{2\beta R_1}\alpha^{-1} . \tag{4.4}$$

The proof of the following estimate will be given later.

$$\int_0^\infty dk \frac{k^2}{|k^2-k_0^2|^2}|c|^2 \leq$$
$$e^{2\beta R_1}\{2^{5/2}\, 3^{-1} t^{\frac{1}{2}}\left(1+\left(\frac{\beta}{\alpha}\right)^{1/4}\right)^2\left(1+t\Gamma\frac{3}{32}\frac{\beta}{\alpha}\right)+4\,\alpha^{-3/4}\,\beta^{-1/4}\} . \tag{4.5}$$

We apply (4.3)-(4.5) to the right-hand side of (4.2) and prove the Lemma:

$$\| P_{ac}(e^{-itH} f_{R_1} - e^{-itk_0^2} f_{R_2}) \|^2 \leq \frac{2}{\pi} e^{2\beta R_1}$$
$$\{2\alpha^{-\frac{1}{2}} + \{2^{5/2}\, 3^{-1} t^{\frac{1}{2}}\left(1+\left(\frac{\beta}{\alpha}\right)^{1/4}\right)^2\left(1+\frac{3}{32}t\Gamma\frac{\beta}{\alpha}\right)+4\,\alpha^{-3/4}\,\beta^{-1/4}\}^{\frac{1}{2}}\}^2 = \|f_{R_1}\|^2 C_3 .$$

<u>Proof of (4.5)</u>: By inserting $k^2-k_0^2 = 2\alpha(k-\alpha)+(k-\alpha)^2+\beta^2+i\Gamma/2$ and $(k_0-k)2\alpha = -2\alpha(k-\alpha)-i\Gamma/2$ we find (notice the cancellations!) that

$$\left|1-e^{it(k^2-k_0^2)}e^{i(k_0-k)2\alpha t}\right|^2 = 4\sin^2\left(\frac{t}{2}|k-k_0|^2\right) .$$

Now we fix $C, D > 0$, define $E = \max\{0,\alpha-C\}$ and $F = \alpha+\min\{C,D\}$, and proceed, using that $\sin^2 x \leq \min\{x^2,1\}$ for all $x \geq 0$, as follows:

$$\int_0^\infty dk \frac{k^2}{|k^2-k_0^2|^2} \left| 1- e^{it(k^2-k_0^2)} e^{i(k_0-k)2\alpha t} \right|^2$$

$$\leq \int_0^E dk \frac{k^2}{|k^2-\alpha^2|^2} 4 + \int_E^F dk \frac{k^2}{|k^2-k_0^2|^2} t^2 |k-k_0|^4$$

$$+ \int_F^{\alpha+D} dk \frac{k^2}{|k^2-\alpha^2|^2} 4 + \int_{\alpha+D}^\infty dk \frac{k^2}{|k^2-\alpha^2|^2} 4$$

$$\leq \left(\frac{\alpha+D}{2\alpha+D}\right)^2 \{4 \int_0^E dk \frac{1}{(k-\alpha)^2} + 4 \int_F^{\alpha+D} dk \frac{1}{(k-\alpha)^2} + t^2 \int_E^F dk ((k-\alpha)^2 + \beta^2)\}$$

$$+ 4 \int_{\alpha+D}^\infty dk \frac{1}{(k-\alpha)^2} \leq \left(\frac{\alpha+D}{2\alpha+D}\right)^2 \{\frac{4}{C} + \frac{4}{C} + t^2 \frac{2}{3} C^3 + t^2 \beta^2 2C\} + \frac{4}{D}$$

$$\leq (1 + D/\alpha)^2 \{2/C + t^2 C^3/6 + t^2 \beta^2 C/2\} + 4/D .$$

We take $C = \left(\frac{2}{t}\right)^{\frac{1}{2}}$ and $D = \alpha^{3/4} \beta^{1/4}$, multiply by $e^{2\beta R_1}$ and obtain

$$\int_0^\infty dk \frac{k^2}{|k^2-k_0^2|^2} |c|^2$$

$$\leq e^{2\beta R_1} \{ (1 + \left(\frac{\beta}{\alpha}\right)^{1/4})^2 t^{\frac{1}{2}} \{2^{\frac{1}{2}} + 2^{\frac{1}{2}}/3 + t \beta^2 2^{-\frac{1}{2}}\} + 4 \alpha^{-3/4} \beta^{-1/4}\}$$

$$= e^{2\beta R_1} \{2^{5/2} 3^{-1} t^{\frac{1}{2}} (1 + \left(\frac{\beta}{\alpha}\right)^{1/4})^2 (1 + \frac{3}{32} t\Gamma \frac{\beta}{\alpha}) + 4 \alpha^{-3/4} \beta^{-1/4}\} .$$

The proof of (4.5), and hence the Lemma, is complete.

The main result is

Theorem 4.6. For all $t \geq 0$ and $R_1 > R_s$ we have, putting $R_2(t) = 2\alpha t + R_1$, that

$$\| e^{-itH} f_{R_1} - e^{-itk_0^2} f_{R_2(t)} \| \leq \| f_{R_1} \| K_3 ,$$

where

$$K_3^2 = 8\pi 3^{\frac{1}{2}} \{1 + 3^{\frac{1}{2}} 8^{-1} (1 + \left(\frac{\beta}{\alpha}\right)^2)\}^2 \frac{\beta}{\alpha} \exp\{4\pi 3^{-\frac{1}{2}} \frac{\beta}{\alpha}\}$$

$$+ \frac{16}{\pi} \{ \left(\frac{\beta}{\alpha}\right)^{1/4} \{2^{-\frac{1}{2}} 3^{-1} (t\Gamma)^{\frac{1}{2}} (1 + \left(\frac{\beta}{\alpha}\right)^{1/4})^2 (1 + \frac{3}{32} t\Gamma \frac{\beta}{\alpha}) + \left(\frac{\beta}{\alpha}\right)^{1/4}\}^{\frac{1}{2}} + \left(\frac{\beta}{\alpha}\right)^{\frac{1}{2}}\}^2$$

Proof. We use that

$$\| e^{-itH} f_{R_1} - e^{-itk_0^2} f_{R_2(t)} \|^2$$

$$= \| P_{ac}(e^{-itH} f_{R_1} - e^{-itk_0^2} f_{R_2(t)}) \|^2 + \| P_e(e^{-itH} f_{R_1} - e^{-itk_0^2} f_{R_2(t)}) \|^2 .$$

The first term is estimated as in Lemma 4.5, the second as follows:

$$\| P_e(e^{-itH} f_{R_1} - e^{-itk_0^2} f_{R_2(t)}) \|^2$$

$$\leq 2(\| P_e(e^{-itH} f_{R_1}) \|^2 + \| P_e(e^{-itk_0^2} f_{R_2(t)}) \|^2)$$

$$\leq 2(\| f_{R_1} \|^2 + e^{-\Gamma t} \| f_{R_2(t)} \|^2) \, C_2 \qquad \text{(Lemma 4.2)}$$

$$= \| f_{R_1} \|^2 \, 4 \, C_2 \qquad \text{(Proposition 3.1)}$$

We have proved that

$$\| e^{-itH} f_{R_1} - e^{-itk_0^2} f_{R_2(t)} \|^2 \leq (4 \, C_2 + C_3) \ \| f_{R_1} \|^2 .$$

The proof is complete.

Remark 4.7. Due to the very explicit proof of Theorem 4.6 it is tempting to claim that the "form" of the error constant given (cf. (1.1) or (5.2)) is "optimal". At least, concerning the neccessity of some t - dependence, we have the following precise result:

$$(e^{-itH} f_{R_1} , e^{-itk_0^2} f_{R_2(t)}) \to 0 \quad \text{for} \quad t \to \infty . \qquad (4.6)$$

There exists a short proof of (4.6) based on Proposition 4.1 and [4] Remark 3.4.

Corollary 4.8. For all $t \geq 0$ and $R_1 > R_s$

$$(e^{-itH} f_{R_1}, \chi_{(0,R_1)} e^{-itH} f_{R_1}) \|f_{R_1}\|^{-2} = e^{-\Gamma t}(1 + y_2) ,$$

where

$$|y_2| \leq e^{\Gamma t} K_3^2 + 2 e^{\Gamma t/2} K_3 .$$

Proof. See the proof of Corollary 3.5.

5. An application to α - decay.

Within the framework of the α - decay model described in Section 1 we now present a proof of the validity of the exponential law for some time-interval.

We let R_1 be the radius of detection, and assume k_0 is a resonance and that f_{R_1} is the α - particle state at the time $t = 0$. The probability P_t, that the α - particle is detected during the time-interval $(0,t)$, is calculated using Corollary 4.8 (y_2 given there) :

$$P_t = 1 - e^{-\Gamma t}(1 + y_2) \qquad (5.1)$$

If for some "large" time-interval $(0,t_0)$, $|y_2|$ is "small" compared with 1, then (5.1) is precisely the law of exponential decay.

The data in the first two rows in the following Table have been taken from [2].

Table 1.

	Ra C'	Ra A	Ur
lifetime Γ^{-1}	4,4 10^{-8} mi.	4,4 mi.	4,4 10^{15} mi.
speed	1,92 10^{9} cm/s	1,69 10^{9} cm/s	1,4 10^{9} cm/s
$2\beta R_1$, $R_1 = 1m$	2 10^{-2}	2 10^{-10}	3 10^{-25}
Γ / E	3 10^{-17}	4 10^{-25}	6 10^{-40}

In the evaluation of $|y_2|$ we can use $\frac{1}{4}\,\Gamma/E$ instead of the quantity $\frac{\beta}{\alpha}$. Also we remark that for $t\Gamma > 1$ (and $t\Gamma < \frac{\alpha}{\beta}$) , K_3^2 is given by

$$K_3^2 \simeq 2^{7/2}\, 3^{-1}\, \pi^{-1}\, (t\Gamma)^{\frac{1}{2}} \left(\frac{\beta}{\alpha}\right)^{\frac{1}{2}} . \tag{5.2}$$

Using (5.2) we find that $|y_2|$ is smaller than 0,2 or 0,01 for $t \in (0,t_0)$, where $t_0\Gamma$ are given as follows:

Table 2.

	Ra C'	Ra A	Ur
$\lvert y_2\rvert < 0,2$ for $t_0\Gamma =$	13	22	39
$\lvert y_2\rvert < 0,01$ for $t_0\Gamma =$	7	16	33

References:

[1] Gamow, G.: Zur Quantentheorie der Atomkernes, Zeitschrift für Physik 51, 204 - 212 (1928).

[2] Gurney, R. W., Condon, E. U.: Quantum Mechanics and Radioactive Disintegration, Phys. Rev. 33, 127 - 132 (1929).

[3] Newton, R. G.: Scattering Theory of Waves and Particles, Springer-Verlag, Berlin, 1982.

[4] Skibsted, E.: Truncated Gamow functions, α-decay and the exponential law, to appear in Commun. Math. Phys.

[5] Skibsted, E.: Truncated Gamow functions and the exponential decay law, to appear.

[6] Skibsted, E.: On the evolution of two- and three-body resonance states, to appear.